AGRICULTURE ISSUES AND POLICIES

TRAVELLING BACK TO SUSTAINABLE AGRICULTURE IN A BIOECONOMIC WORLD

THE CASE OF ROXBURY FARM CSA

AGRICULTURE ISSUES AND POLICIES

Additional books in this series can be found on Nova's website under the Series tab.

Additional e-books in this series can be found on Nova's website under the e-book tab.

AGRICULTURE ISSUES AND POLICIES

TRAVELLING BACK TO SUSTAINABLE AGRICULTURE IN A BIOECONOMIC WORLD

THE CASE OF ROXBURY FARM CSA

JOHN M. POLIMENI
RALUCA-IOANA IORGULESCU
AND
RICHARD SHIREY

Library of Congress Cataloging-in-Publication Data

Travelling back to sustainable agriculture in a bioeconomic world: the case of Roxbury Farm CSA / editors: John M. Polimeni, Raluca Ioana Iorgulescu and Richard Shirey (Albany College of Pharmacy and Health Sciences, NY and others).
pages cm. -- (Agriculture issues and policies)
Includes index.
ISBN 978-1-63463-376-5 (hardcover)
1. Sustainable agriculture--United States--Case studies. 2. Community-supported agriculture--United States--Case studies. I. Polimeni, John M., editor. II. Series: Agriculture issues and policies series.
S441.T73 2015
631.50973--dc23
2014040969

Published by Nova Science Publishers, Inc. † New York

CONTENTS

List of Figures

LIST OF TABLES

PREFACE

In our world ravaged by increasingly extreme environmental phenomena, *bioeconomy* is one of the concepts combining the sustainable production of renewable resources from land and water and their conversion into indispensable goods and energy. Nicholas Georgescu-Roegen introduced, in the late 1960s, research on *bioeconomics* as he foresaw the conflict between humans and the environment, especially in the loss of soil fertility for agriculture. This book uses Georgescu-Roegen's framework to focus on the connection between agriculture and environment and will provide an in-depth examination of one form of sustainable agriculture that can potentially result in economic growth, namely, community supported agriculture (CSA).

The opening chapter follows *The Trip Back to Sustainable Agriculture* looking at the changes of agricultural systems in developed countries, from traditional agriculture to industrial agriculture and striving to get back to regain what was almost lost in the process: a sustainable agricultural system based on ancient, traditional know-how adapted to different areas of the world, sometimes under extreme climate conditions. We believe that this know-how, wisely 'transplanted' to regions that are suffering from climate change effects, could provide food security and food safety.

The topic of food security is further debated in the chapter *Sustainable Agriculture: Economic Growth and Food Security Tradeoff.* Here a relatively simple description of why agriculture is a vital economic sector, how sustainable agricultural approaches can be used as an economic development tool and the relationship to food security are introduced; all in the bioeconomic context of economic growth in a balanced ecosystem. We thoroughly describe how agriculture is an important economic sector to an economy and explore how sustainable agriculture can be used to grow an economy in both developed and developing economies. Finally, we examine why developing the agricultural sector for any nation is important for food security.

In our time, one of the most silent and dangerous environmental threats is the loss of soil, the most important element in feeding humanity. The soil erosion and desertification that affect more and more areas of the Earth are directly related to modern, industrial agricultural practices. The chapter on *The Threat of Soil Erosion and Sustainable Agriculture* offers a quick excursion through this 'fearful land' and makes the case for an urgent *green* Green Revolution.

The chapter *Community Supported Agriculture: What Is It?* provides a history of CSA, exploring its beginnings and how it has evolved over the years. In particular, we will clearly explain the different approaches to CSA from around the world; some CSA operations have livestock, others only vegetables, etc. Furthermore, this chapter will plainly state what CSA is and its benefits and present a theoretical concept of supply and demand of CSA.

The *History of Roxbury Farm CSA* opens the case study of Roxbury Farm, which will cover three chapters. Roxbury Farm is one of the largest CSA operations in the United States. The history of the farm and how the farm has evolved over time are discussed in this opening chapter. Furthermore, information on the operations of the farm will be provided since the way the farm is managed is one of the key aspects of Roxbury Farm's success.

The Financial History of Roxbury Farm: 2000 – Present complements the history and operations of Roxbury Farm by presenting the budget of the farm and detailing how the budget has changed as the farm has evolved. This chapter builds upon the chapter *Community Supported Agriculture: What Is It?* to examine the operating costs and revenues of the farm in order to provide more details on how a successful CSA operates and how CSA methods can be applied to other regions of the world. This information is important because it can be used by other CSA operations or farms considering becoming a CSA business.

Additional details about farm operation and its connection to members is presented in *Roxbury Farm Annual Survey Results: 2003-2009*. The survey questions, administered anonymously to Roxbury Farm members since 2003, explore a variety of topics, such as why people join Roxbury Farm, the perceived benefits on human and environmental health, assisting local farmer, and how CSA farms affect food security. This chapter is important because how a CSA approach can be used as a rural economic development tool is described, raising both incomes and employment in rural regions possibly stabilizing these areas. The resulting rural poverty reduction from better conditions in those areas has been an engine in overall poverty reduction.

The *Concluding Comments* wrap-up the book by reiterating the overall environmental, economic, and social benefits of CSA in relation to a discussion of public-policy issues related to agriculture. We discuss the issue of land tenure and how a CSA approach can be used to improve rural economic development. Within this context food security is also examined. The use of agricultural subsidies by Western countries is part of this discussion because these subsidies have a major negative impact on farmers, particularly those in developing countries. Agricultural policies in the developed world have resulted in lower production prices which contribute to increased poverty among the rural poor in developing countries. We detail how CSAs can offset these negative effects. Furthermore, once the environmental and public health costs, such as fertilizers, hybrid seeds, pesticides, irrigation, incidences of cancer, are included in the real costs of industrialized agriculture, food is not cheap.

As consumers consider these costs in their food purchases, or if food costs continue to increase with the demand for biofuels, the price a consumer pays in the store for industrially produced food is quickly converging with that of food produced in a sustainable manner. Thus, a sustainable approach to agriculture, such as CSA, is highly cost-competitive when all costs are internalized to the farm.

CSA has garnered significant interest in the past decade as more consumers understand the problems with food produced in a conventional, industrial manner. While other books and research articles have explored CSA and CSA operations, few contain the theoretical background and data that this book will present. To the best of our knowledge the dataset used in this book is the longest and, perhaps, most comprehensive survey of any CSA in existence. As a result, we hope to provide the reader with meaningful insight about CSA, the problems CSA operations have, and how CSA farms can evolve over time.

ACKNOWLEDGMENTS

Our gratitude goes to the farmers at Roxbury Farm CSA, Jean-Paul Courtens and Jody Bolluyt, and to all the farm members who gave their support for this research.

The authors would like to thank those who reviewed previous versions of this book and provided invaluable suggestions. All remaining mistakes are ours. The authors worked equally on the contents of the book and the views expressed within are solely the responsibility of the authors and should not be interpreted as reflecting the views of Roxbury Farm CSA or its members.

We are grateful to our families for their patience and support over all the years of work on this book.

John M. Polimeni and Raluca-Ioana Iorgulescu dedicate this book to the loving memory of John Joseph Polimeni, whose love and kind encouragement provided the strength necessary to finalize this work.

About the Authors

John M. Polimeni is an Associate Professor of Economics and Director of Graduate Studies in Health Outcomes Research at the Albany College of Pharmacy and Health Sciences, USA and is the author of *"Jevons' Paradox and the Myth of Resource Efficiency Improvements"*, *"The Economic Growth, Environment, Public Health Connection: An Ecological Economic Case Study of the Impact of the Yali Falls Dam on Cambodia"*, and *"Spatial Simulation Modeling: Projecting Residential Development in the Hudson River Valley of New York State"*.

Raluca-Ioana Iorgulescu is a senior researcher with the Institute for Economic Forecasting-NIER, Romanian Academy, Romania, and an associated senior research economist with Resource Dimensions, USA; her work relates to sustainable development, multi-scale integrated assessment, socioeconomic and ecological metabolism, and sustainable agriculture. She holds a B.S. in Physics, B.A. and Master's in Economics and a Ph.D. in Ecological economics.

Richard L. Shirey is a Professor Emeritus of Economics at Siena College. He has been active in support of the Roxbury Community Supported Agriculture for many years. His participation at Roxbury has been multifaceted: as advisor, worker, fund-raiser, member, and consumer. His fascination with biodynamic agriculture derives from its positive attributes including sustainability, building the human spirit and renewing the planet.

Chapter 1

The Trip Back to Sustainable Agriculture

Abstract

Agriculture is complex, on one hand providing food for the world while on the other hand bringing considerable environmental degradation. The balance of this give and take relationship, by many people's standards, has gone from positive to negative, largely due to the multitude of environmental and population issues coming to the forefront. As a result, agricultural production has undergone considerable scrutiny resulting in strong consumer movements for sustainable agriculture. However, sustainable agriculture is often misunderstood as to what it is. This chapter presents an overview of the problem of industrial agriculture, provides our definition of sustainable agriculture and, also, the purpose of this book.

1.1. Overview of the Problem

Dare we hope for a better future on a safe and sustainable planet where all of the world's people are healthy and happy? The planets' life-support systems are increasingly stressed by industrial and agricultural practices that exploit nature while introducing artifacts unknown to nature that are eroding the carrying capacity of the planet into the foreseeable future. What do we have to look forward to over the next century as the world's population faces ecological challenges that are overwhelming and seemingly intractable and where a principal engine of degradation is the world's industrial food system? These systems are embedded into the institutional fabric of the political and economic systems of Western developed countries. They deliver most of the raw and processed food that is consumed by the majority of Westerners, as well as many millions around the planet every day. The Earth's natural resources are being depleted at an unsustainable rate. Clearly, the agricultural systems of the future must be different. Since these systems are created by humans and can be changed by them, what kind of change will we foster?

These challenges cannot be answered with military force, such as the responses in World War I and II, nor will they be resolved with an overwhelming response of money poured to existing institutions, as with the bank bailouts in 2008 and 2009. However, a large and immediate response is needed on both the local and global levels because the world is rapidly moving toward an ecological catastrophe, one that is a result of human activity and for which the environmental, political, and socially disruptive consequences can only continue to be ignored at our peril.

The major ongoing and worsening ecological problems related to the industrial food system that the world's people must address include environmental degradation in the form of soil depletion, river and estuary pollution, fresh water shortages, climate change in the form of temperature and weather extremes, the loss of plant and animal diversity, and the impact of overpopulation.

These problems are systemic and interrelated. They have an impact on and are impacted by one common overlying factor, the world's industrial agricultural system. Ironically, agricultural activity that has been so productive has also degraded and destroyed so much of the life support capacity of the planet. However, with the proper changes into the future, agriculture can also be the salvation since agricultural activity can have a positive effect on natural systems. Major changes, some would say radical changes, must be made in agricultural practices and food distribution systems if the deleterious momentum of the industrial agricultural system is to be channeled into a direction more harmonious with sustainability.

We assert that from seed to table, sustainable and vibrant local agricultural systems embedded in the local ecosystem are vital for human health and well-being, for healing the planet, and for contributing significantly to social and political stability. In the coming years the agricultural system will be called upon to feed a projected nine billion people by the year 2050 (United Nations, 2005a; 2005b; 2011) while competing for ever-decreasing freshwater, land, and energy resources. However, the expansion of current industrial agricultural practices will only exacerbate environmental problems further. Hence, a new agricultural system must begin to heal the planet of the most egregious impacts of current production and make food available to people where they live. Rather than competing for more natural resources, agricultural practices could be contributing to the availability of more healthy land, fresh water and breathable air.

1.2. THE ROAD TO INDUSTRIAL AGRICULTURE

Due to the availability of farmland and market incentives, American farmers, in general, have typically responded to the demand of an increased population by expanding the acreage of the land that was cultivated, narrowing the variety of crops produced, and increasing the use of external inputs, such as fertilizer, in an attempt to increase the output per acre. For example, agriculture expanded across the North America n continent during the westward expansion of the 19th century by putting much more land into cultivation. The expansion was accompanied by a myriad of technological changes such as the combine and the reaper. As farmers turned to 20th century technological innovations to expand their production, the amount of land that a farmer could effectively farm also increased dramatically, but not without soil depletion and erosion, as evidenced by the severity of the Dust Bowl of the 1930s, by ripping up the prairie and exposing vast swaths of open land to wind. Agricultural tilling practices have since improved to minimize erosion from the wind, but little has been done to restore the vitality of the soil.

By the mid-20th century, the introduction of manmade synthetic chemicals, often made from fossil fuels, boosted agricultural production tremendously. Consequently, for nations of the Industrial world, the Malthusian specter of population growth outstripping the growth of the food supply was held at bay, although food shortages continue to stalk many poorer nations, a situation compounded by the loss of traditional small farms and farmers in many of those nations.

The Haber-Bosch process, originally developed for military purposes, was found to be effective for creating nitrogen-based chemical fertilizers (Smil, 1997). Hence, in addition to the mechanical advances that began to significantly reduce labor inputs to farming in the 19th century, the 20th century Industrial Revolution introduced chemical pesticides and fertilizers into agricultural production. During this time, the mechanization of plowing, planting, cultivation, and harvesting in agricultural production continued, but with increasingly more powerful fossil fuel powered tractors. These technologies, along with supportive agricultural policies, enabled farmers who adopted the new technologies, to cultivate more land and produce more food per hectare. Average farm size increased dramatically while the number of farmers diminished greatly.

The limiting factor of soil nitrogen, which formerly depended on natural soil renewal and animal manure, was no longer a problem (Sachs, 2008: 65) because it was now provided by chemical fertilizers. Not only are these practices depleting soil, but they are dependent on vast amounts of fossil fuels to produce the chemicals, apply chemical inputs to the fields, and to harvest and distribute the crops. The by-products of the process are more air, ground and water pollution on a vast scale.

Eventually the potential land available for further agricultural production was already exploited, depleted or converted into residential or commercial properties. Furthermore, the introduction of chemicals depleted the soil of its ability to support the microbial life needed to grow healthy plants. In addition, many farmers were no better off despite the increased production because of the need to purchase ever increasing amounts of inputs for their farms while having to compete in very competitive markets. Ironically, in spite of the increase in productivity at the expense of the environment, the farmers themselves suffered from decreasing profits that they could earn over the past century because their costs for external farm inputs and other supplies, food processors and distributors have increased.

Additionally, the Green Revolution, which arguably hit its peak in the 1960s, provided the technological advancements that allowed marginal and depleted lands to be farmed and existing agricultural land to be farmed more productively. However, over the years the Green Revolution has not delivered on its promise. Although these innovations resulted in hybrid plants that selectively resisted drought, pests, and weeds and, as a result, brought the promise of more food production, they had their own weaknesses such as requiring more fertilizer because more output was taken from the land. Furthermore, the hybrid plants required more pesticides because of the capacity of insects to develop resistance to insecticides. In the end, the increased output was cancelled out by increased input costs and environmental damage. However, the seed and chemical industry prospered. This industrial agricultural model further promoted recent technological developments, such as genetically modified seeds and stronger nitrogen-based fertilizers. Over time, however, the use of these new technologies required increasing levels of fertilizer applications to maintain production levels.

During the 1970s, then Secretary of Agriculture Earl Butz transformed U.S. government agricultural policies from a crop price subsidy program that encouraged the limitation of agricultural production to keep commodity prices high, to a strategy that used subsidies to encourage increased production of commodity crops that resulted in lower world food prices. This change in policy resulted in farmers planting as much of their land as possible and forced many other countries to follow suit with agricultural subsidies, particularly in Western Europe and Japan. As a result, the farmers in those countries took similar approaches to their farming methods. The encouragement for additional production along with all the aforementioned forces of change resulted in the industrial agriculture system that continues to dominate the developed world today.

1.3. Some Issues Associated with Industrial Agriculture

These technologies and policies have led to a *drastic reduction in the number of small farms* because the costs of scaling up their production to take advantage of the subsidies are prohibitive. As a result, many small farms were sold or went out of business because changes in farm policy and technology led to a drastic decrease in income for farmers who could not acquire more land or machinery. For example, the average farmer's income in the U.S. has fallen 32% since 1950, though the average farm size has greatly increased, leading to a highly concentrated agricultural system where the largest farms receive the largest subsidies and profit. Moreover, for every dollar spent on agricultural products, farmers today receive 10 cents or less as compared to nearly 70 cents they received during the 1950s (Spector, 2002: 288-289). In effect, Western governments' subsidies have made the wealthy farmers even wealthier as approximately 70% of the payments go to the top 10% of the farms by gross income in both the United States and Europe (Henderson and Van En, 2007: 17).

The agricultural system worldwide continues to transform from decentralized-small farms providing subsistence for farm families and neighbors to a *centralized industrial-large farm approach*. One consequence of this transformation has been the rapid growth of city populations in many lesser developed nations. In addition to the migration problems, the changes in the agricultural system have created an entirely new set of problems that must be addressed. We will briefly examine what we think are some of the most important issues associated with industrial agriculture and the technological changes accompanying this transformation that have affected the health of the environment and, ultimately, human health.

Industrial agriculture has led to a dramatic *transformation of the landscape of the planet*. The change in the landscape has come from two extremely destructive activities: turning forests into farmland and paving over prime farmland for residential, commercial, and urban expansion. In many regions, particularly those in developing countries, massive deforestation has generally resulted from harvesting trees to sell logs for lumber. Once the land is stripped of canopy, it is subject to erosion and depletion of residual nutrients. As a consequence, the ability of nature to convert carbon dioxide into oxygen is greatly reduced. Moreover, with more powerful tractors and more potent fertilizers, farmers are able to grow additional crops further depleting the soil. These activities not only tend to move farmers off the land but they also reduce the diversity of plant and animal life as well as contribute to the degradation of land as compared to the formerly forested areas. Moreover, as previously discussed, the biotechnological revolution has led to genetically modified seeds (GMOs), fertilizers, and pesticides that, in combination with the expansion of irrigation technologies, have enabled farmers to grow crops on land previously considered unsuitable for agricultural purposes, but at great environmental cost.

At the same time, *urban encroachment* in different regions of the world has taken much of the prime farmland out of production because land values have risen so much that many farmers have sold their land for more than any prospective farmer would be willing to pay since the agricultural value is tied to the productive capacity of the land. Industrial agricultural policy has overwhelmingly favored the large farm activity at the expense of small farmers even though the productivity of small farms can rival that of their industrial counterparts. Furthermore, several other forces are now operating to reduce the viability of the small farmer. These include urban sprawl and government support for the production of biofuels, which has caused high production costs which are often too great of a barrier for small farms to overcome.

One of the outcomes of subsidizing the industrial food system by Western nations has been that *food has been travelling farther distances* to reach consumers, making it less fresh and nutritious, consuming even more fossil fuels and contributing to increased air pollution levels. For example, the globalization of food has resulted in food traveling an additional distance of

50% for the United Kingdom between 1978 and 1999 (Pfeiffer, 2006: 25). In the United States, food travels approximately 1,600 to 1,800 miles from farm to plate. In addition, many exotic foods and flowers travel by air in air-conditioned comfort. Some would argue that the result is good for the balance of trade for the nations that export food, but long distance food exchange is not environmentally sustainable nor is it in the best interest of consumers who have no access to the place where their food is grown. Thus, many do not have assurance that the food they are consuming is safe.

Unfortunately, the changes in the landscape and the globalization of many elements of the food supply have contributed an *increasing amount of carbon dioxide, CO_2,* a greenhouse gas, to the atmosphere. In addition, the industrial agricultural system is responsible for significant emission levels of methane, nitrous oxides and sulfur hexafluoride, all of which have a larger warming effect per unit than carbon dioxide (Lappé, 2010: 7-9). Besides the use of fossil fuels by tractors and other farm machines which are obvious CO_2 sources, industrial agriculture is one of the most energy intensive, CO_2 polluting sectors. The chemical fertilizers and pesticides prevalent in today's agricultural system are fossil-fuel based and require a large amount of energy to produce. Additionally, pumps that are used to ensure crops receive water through modern irrigation systems are fossil-fueled. Besides chemicals and irrigation, the energy used to process, distribute, and prepare food greatly surpasses the energy consumed to produce food (Pimentel, et al., 1989; Hendrickson, 2008) making food production and consumption a contributor to global climate change.

The decline in small farm income and urban expansionary pressures have led small farmers to sell their land for residential and commercial development, while at the same time, many larger farms have merged into mega-farms. Ironically, the energy crisis has accelerated landscape change as government policy increasingly promotes renewable forms of energy, such as *biofuels*. The increased emphasis on biofuels makes monocropping, already rampant in industrial agriculture, worse as land was shifted away from food production (for animal or human consumption).

For example, the purpose of the biofuels policy in the United States is to reduce the nation's dependency on imported foreign oil by substituting domestically produced biofuels for some fraction of imported oil. Most of the biofuel in the US is made from corn and processed to be added to gasoline for use in motor vehicles. While this policy addresses the fossil fuel dependency problem, it introduces an array of additional problems equal to or greater than the original concern. For example, some farmers can receive higher prices for biofuel producing crops, mostly corn. The result, however, is that only corn is now grown on land that was once used to grow different crops thereby depleting the soil even more rapidly than it would have been. Less land is devoted to human food production.

Overall, there is an upward pressure on food prices because more of the agricultural production is transferred for purposes other than for human consumption. The result may be for Americans to spend a percent or more on food than they are accustomed to, but higher food prices have a devastating effect on consumers in poor countries who are dependent on imports of food from the US for their survival. Within the past few years, biofuel production has been responsible for a 10% to 30% increase in world commodity prices (Peterson, 2009: xix).

1.4. WATER IMPACT OF INDUSTRIAL AGRICULTURE

Klepper (1979) argued that increasing levels of CO_2 emissions, thought by many scientists to cause higher global temperatures and to disrupt the hydrological cycle, have a significant effect on world water shortages and desertification. As climate change occurs one of the consequences has been extreme weather anomalies such as frequent flooding and stronger storms.

Furthermore, in the past several decades a large number of hydroelectric dams have been constructed around the world for energy development and irrigation purposes. However, these hydroelectric dams have exacerbated water shortages and caused a tremendous amount of landscape transformation, contributing factors for flooding, mudslides, and earthquakes. The impact has been on both upstream and downstream communities as the ecological and hydrological systems are altered which negatively affects agriculture by eliminating fertile land, as well as hurting those farms that are dependent upon irrigation. Many dams have also been filling up with silt which defeats both their hydroelectric potential and available water for irrigation.

In addition to water shortages caused by increasing consumption and a disrupted hydrological cycle, much of the remaining water is contaminated at an increasing rate by pollution from farms. The monocropping prevalent in today's industrial agriculture is partially responsible for the nitrogen in the groundwater. For example, soybeans, a heavily subsidized crop in the United States, have root systems that contain nitrogen-fixing bacteria that add nitrogen to the already overloaded soil when the soil is treated with chemical herbicides, nitric oxide (NO) and nitrogen dioxide (NO_2); the excess emissions from herbicide-treated soybean plants seep into the groundwater (Klepper, 1979).

Moreover, animal manure, typically concentrated into small areas in the industrial agriculture system within confined animal feeding operations (CAFOs) regularly contaminates the groundwater and nearby streams, especially during severe storms and flooding. There have been occasions where manure stored on farms, for example the hog farm incidents in North Carolina in 1995 and 1999, has leaked into a nearby river or stream, severely contaminating the water supply (The New York Times, 1995; Howlett, 1997). In fact, agriculture has accounted for more than 70% of the pollution in rivers and streams in the United States (Henderson and Van En, 2007: 14).

The pollution from animal waste is worsened because the policy of the industrial agricultural system is to routinely inject animals with large amounts of antibiotics and hormones to make them grow larger and faster while fighting-off disease. Corn-fed beef cattle are especially susceptible to infections because their intestinal systems are not designed to digest corn very well. As a result, almost all beef cattle in CAFOs are given antibiotics as a prophylactic measure. The confinement itself, whether for beef, pigs, chickens and even fish, tends to increase the rate of bacterial infection. These antibiotics and hormones then get into the water system potentially consumed by humans. They also alter the soil and water composition as well as affect region's flora and fauna.

Additional pollutants come from the nitrogen-based chemical fertilizers and pesticides that are used to increase production (see the previous example). The nitrogen, in addition to tainting agricultural output eventually seeps into the groundwater, flowing into rivers and streams and contaminating the water. Estimates find that only 18% of all nitrogen compounds applied to farm crops are absorbed by the plants (McKenney, 2002: 126).

1.5. Effects on Human Health of Industrial Agriculture

The U.S. Food and Drug Administration (FDA) found that at least 53 pesticides applied to food crops are carcinogenic (Kimbrell, 2002: 10), whereas the U.S. Environmental Protection Agency (EPA) has identified more than 165 pesticides as potentially carcinogenic (Kimbrell, 2002: 11). When ingested by humans, these nitrogen-based compounds can lead to methemoglobinemia (McKenney, 2002: 127). If the cost of eliminating or mitigating the effects of these chemicals were to be internalized by the farmers who use them, there would be an incentive for the farmers to reduce or eliminate using them while seeking non-polluting alternative agricultural practices.

In addition to the carcinogenic impacts, the exposure to hormones and antibiotics leads to antibiotic resistance, which in turn results in ordinary bacterial infections in humans that are much more difficult to effectively treat. The Centers for Disease Control (CDC) found that between 1970 and 1999, food-borne illnesses increased more than ten times, or approximately 80 million people a year in the United States alone (Kimbrell, 2002: 10, 12). The World Health Organization (WHO) estimates that the food supply poisons more than 25 million people each year.

As can be seen, the evolution of technological changes and policies related to industrial agriculture have had and continue to have a large negative impact on environmental and human health. However, despite the problems like those listed above, policy-makers continue to turn to science and technology to boost agricultural output and offer no more than a technological fix for the problems discussed. The belief in technological solutions has arisen concurrently in the context of large monocropping farms that themselves are reliant on very specialized agricultural technologies.

There is growing awareness that the methods and practices used in industrial agriculture cause a large amount of pollution and are both fool-hearty and unsustainable. By no means are we saying all technology is bad. Rather, we suggest that the solution should be to use technology with sustainable agriculture approaches that reverse the damage to the environment and to people. Make no mistake; this approach to agriculture will be a difficult transformation. What could be easier than the industrial agricultural method of using heavy machinery and applying chemicals to grow one or two crop varieties while ignoring the long-term social and environmental costs; in effect keeping the private costs of production down by transferring the social costs to neighbors and the environment?

However, as will be shown in this book, sustainable agriculture systems are necessary. Before going further we will define what sustainable agriculture means in regards to this book.

1.6. THE NEED FOR SUSTAINABLE AGRICULTURE

The term 'sustainable agriculture' tends to illicit an emotional response because the term means something different to various groups of people. One group believes that sustainable agriculture is unrealistic and that the problems of industrial agriculture are overstated, that the current practices only have to be adjusted slightly to feed the world. Proponents of sustainable agriculture believe that the transition to sustainability is needed soon, that industrial agriculture is the problem, and that the social and environmental costs of agriculture need to be internalized. Clearly, sustainable agriculture is an enemy to some and a friend to others.

However, while controversial, the term is the consensus choice on how to describe agricultural practices that differ from the standard, modern, industrial approach (Bidwell, 1986; World Bank, 1981). According to Gips (1987) the most common definition of sustainable agriculture is an agricultural approach that is ecologically sound, socially just and humane, and viable from an economic perspective. For our purposes we define sustainable agriculture as being economically viable and socially just, while minimizing the use of chemicals, such as fertilizers, pesticides, or any other external inputs; ideally eliminating them completely. The farm is treated as a whole system achieving viable, healthy production levels while maintaining the stability of the ecosystem in which the farm is located, and which is made available to future generations of farmers.

Thus, sustainable agriculture must be, by its very nature, diverse and complex. Sustainable agriculture preserves biodiversity, soil fertility, and water purity, recycles natural resources and uses locally available renewable resources when available, and conserves energy. Such an approach increases self-sufficiency, provides a stable income in rural areas, and respects the diversity and interdependence of the ecosystem by using science and technology to complement the wisdom accumulated over centuries by farmers around the world (Henderson and Van En, 2007: 23). Farmers using sustainable agricultural techniques must understand and monitor the complex relationships that exist on each individual farm, such as the association among and between plants, animals, and the soil, so these interactions work in conjunction with each other to stay in balance. Common practices include composting, crop rotation, and intercropping while using organic fertilizers and pesticides and avoiding the use of growth hormones and anti-bacterials.

We describe some of the more common terms associated with sustainable agriculture in Chapter 3 in which we also advocate a transformation of agriculture and agricultural policy to encourage the growth of small local farms that produce primarily for local markets. This move would encourage local, rural economic development as consumer purchases and additional farm purchases resulting from increased sales would lead to additional employment and more local sales tax, keeping money in the region. Consumers would also have access to the source of their food as well as effective ways to actively monitor food safety.

Furthermore, a move to local, sustainable agriculture and a change in agricultural policy would result in less environmental degradation as fossil fuel consumption expended to transport food to consumers would decrease.

1.7. Purpose of the Book

The purpose of this book is to illustrate how sustainable agriculture, especially the expansion and use of local, small-scale farms, can contribute to an increase in food security and to a revitalization of rural economic development, which is discussed in detail in Chapter 2. For example, in the United States less than 5% of food is produced locally (Pfeiffer, 2006: 68). Furthermore, a 2010 U.S. Department of Agriculture report revealed that are 17.2 million households in the United States that are food insecure (Coleman-Jensen et al., 2011). If a transformation were to occur to where just 10% of food was produced locally, the economic and environmental impact would be substantial.

Envision the impact if a local agriculture movement were to happen on a wide-scale. Sustainable farms located near population centers where most of the food insecurity occurs could be part of the solution to the problem as well as helping to reduce the public health problems that many urban dwellers have from poor food choices. Furthermore, since these farmers would purchase their inputs from local businesses and potentially provide products for food-processing plants, they could provide a large economic multiplier effect in their local economies (Hinrichs, 2007: 29). The dollars the farmers spend locally circulate several times more through the local economy than the money spent by multinational corporations (Lyson, 2004: 62). The health of nonfarm economic activity (i.e., health care, education, banks, and other services) in rural communities is highly correlated to the numbers and types of farms in a region (Ikerd, 2008: 141).

This book will focus on Community Supported Agriculture, or CSA for short. CSA is particularly attractive for farmers and a widely used option throughout the world. Typically, CSA operations will use sustainable agricultural approaches to produces food. CSA farms function by selling shares to consumers (i.e., members); in exchange, members receive a regular delivery of a variety of goods produced by the farm during the growing season. CSA is a quasi-market structure in which the members actively support the farm and the farmers through some kind of nonmarket activity beyond the purchase of a membership. Therefore, a human relationship beyond the market relationship is formed between the farmer and the members; more detail is provided in Chapter 4. In this book (Chapters 5-7) we present the case study of Roxbury Farm CSA to illustrate the effectiveness of sustainable small-scale farming.

Roxbury Farm is located on a 450 acres plot of land, of which slightly less than 100 acres are farmed to grow vegetables, in Kinderhook, N.Y. Roxbury is one of the largest CSA farms in the United States with about 1,100 members. Survey data from Roxbury Farm members were collected annually from 2003 to 2009 on a variety of subjects related to the farm. Members were asked, among other things, questions about: 1) their socioeconomic status; 2) what factors were important to them when they joined the farm originally; 3) what factors are important to them after at least one year of membership; 4) their satisfaction with product quantity, quality, variety, appearance, and taste; 5) how their eating habits have changed from before they were members; 6) their position on environmental issues; 7) their viewpoints on health issues; and 8) their stance on food security issues. We will illustrate, using the data collected from the surveys, how CSA can be used to increase food security while improving environmental health, human health and rural economic conditions.

Thus, the CSA approach is fundamentally different from today's conventional agriculture which promotes mega-farms and monocropping in the name of efficiency and production for the market. CSA farmers are producing for members with whom they have more than a market relationship. In the standard neoclassical economic approach, the conventional agricultural model has been one where large farms capture economies of scale producing goods more cheaply than small farms. External costs to the environment are not included in the calculation of farm cost unless regulation forces those costs to be internalized.

The products from large farms flow into the market by linkage with large wholesalers, who then link with large retailers, who then serve the mass market (Hinrichs, 2007: 24). However, as will be seen throughout this book, these economies of scale and long distance transportation to markets come at a very high, though sometimes invisible cost.

On the other hand, sustainable agriculture can produce an average 93% increase in food production per hectare and require half of the energy input per unit area that industrial crops require (Pfeiffer, 2006: 68).

This book will show how the reliance on industrial agriculture is reckless; destroying the environment while also being fiscally unsustainable without a heavy dependence on subsidies for its continuation. A valid alternative, when sustainable farming methods are used, is CSA. The book is organized so as to provide a thorough introduction of CSA from its origins, through a case study of Roxbury Farm, to the applications of CSA in other regions of the world as a possible method of economic development.

References

Bidwell, O.W. (1986). Where Do We Stand On Sustainable Agriculture? *Journal of Soil and Water Conservation, 41*(5), 317-329.

Coleman-Jensen, A., Nord, M., Andrews, M., & Carlson, S. (2012). Household Food Security in the United States in 2011. *ERR-141*, U.S. Department of Agriculture, Economic Research Service, September.

Gips, T. (1987). *Breaking the Pesticide Habit: Alternatives to Twelve Hazardous Pesticides.* Minneapolis, MN: International Alliance for Sustainable Agriculture.

Henderson, E. & Van En, R. (2007). *Sharing the Harvest: A Citizen's Guide to Community Supported Agriculture.* White River Junction, Vermont: Chelsea Green Publishing Company.

Hendrickson, J. (2008). Energy Use in the U.S. Food System: A Summary of Existing Research and Analysis. *Center for Integrated Agricultural Systems Working Papers Series.* http://www.cias.wisc.edu/wp-content/uploads/2008/07/energyuse.pdf

Hinrichs, C. C. (2007). Practice and Place in Remaking the Food System. In C. Clare Hinrichs & Thomas A. Lyson (Eds.), *Remaking the North America n Food System: Strategies for Sustainability* (pp. 1-15). Lincoln and London: University of Nebraska Press.

Howlett, D. (1997). Lakes of Animal Waste Pose Environmental Risk. *USA Today*. December 30.

Ikerd, J. E. (2008). *Crisis & Opportunity: Sustainability in American Agriculture.* Lincoln and London: University of Nebraska Press.

Kimbrell, A. (2002). Myth Two Industrial Food is Safe, Healthy, and Nutritious. In A. Kimbrell (Ed.), *The Fatal Harvest Reader: The Tragedy of Industrial Agriculture* (pp. 10-14). Washington: Island Press.

Klepper, L. (1979). Nitric Oxide (NO) and Nitrogen Dioxide (NO_2) Emissions from Herbicide-Treated Soybean Plants. *Atmospheric Environment, 13* (4), 537-542.

Lappé, A. (2010). *Diet for a hot planet: the climate crisis at the end of your fork and what you can do about it.* Bloomsbury Publishing USA.

Lyson, T. A. (2004). *Civic Agriculture: Reconnecting Farm, Food, and Community.* Medford, Massachusetts: Tufts University Press.

McKenney, J. (2002). Artifical Fertility: The Environmental Costs of Industrialized Fertilizers. In A. Kimbrell (Ed.), *The Fatal Harvest Reader: The Tragedy of Industrial Agriculture* (pp. 121-129). Washington: Island Press.

Peterson, E. W. F. (2009). *A Billion Dollars a Day: The Economics and Politics of Agricultural Subsidies.* West Sussex: John Wiley & Sons Ltd.

Pfeiffer, D. A. (2006). *Eating Fossil Fuels: Oil, Food and the Coming Crisis in Agriculture.* Gabriola Island, BC: New Society Publishers.

Pimentel, D., Armstrong, L. Flass, C. Hopf, F. Landy, R. & Pimentel, M. (1989). Interdependence of Food and Natural Resources. In: D. Pimentel & C. Hall (Eds.) *Food and Natural Resources.* San Diego, CA: Academic Press, Inc.

Sachs, J. D. (2008). *Common Wealth: Economics for a Crowded Planet.* New York: The Penguin Press.

Smil, V. (1997). Global Population and the Nitrogen Cycle. *Scientific American.* July: 76-81.

Spector, R. (2002). Fully Integrated Food Systems: Regaining Connections between Farmers and Consumers. In A. Kimbrell (Ed.), *The Fatal Harvest Reader: The Tragedy of Industrial Agriculture* (pp. 288-294). Washington: Island Press.

The New York Times. (1995). Huge Spill of Hog Waste Fuels an Old Debate in North Carolina. June 25.

United Nations. (2005a). World Population 2004. *Department of Economic and Social Affairs, Population Division,* August.

United Nations. (2005b). World Population to Increase by 2.6 Billion over Next 45 Years, With All Growth Occurring in Less Developed Regions. Press Release Pop/918, February 24.

United Nations. (2011). World Population Prospects: The 2010 Revision. *Department of Economic and Social Affairs, Population Division.*

World Bank. (1981). *Principles and Technologies of Sustainable Agriculture in Tropical Areas: A Preliminary Assessment*. Washington, D.C.: The World Bank.
World Health Organization. http://www.who.int/en/

Chapter 2

SUSTAINABLE AGRICULTURE: ECONOMIC GROWTH AND FOOD SECURITY TRADEOFF[1]

ABSTRACT

Many countries, particularly developing countries, are not food secure. There are a number of reasons for this problem, chief among them agricultural subsidies provided to farm operations in developed countries. As a result, many developing countries do not produce enough food to be secure because farmers cannot compete with the artificially low prices of food from developed countries due to the subsidies. However, this trend is unlikely to continue as farm operations in developed countries must increase the amount of inputs, such as fertilizer, to maintain their levels of production. Furthermore, the economic realities of large national debts will result in budget cuts, of which agricultural subsidies will be a likely target. Moreover, consumers are increasing their demand for food grown sustainably. Therefore, an opportunity exists for rural regions and developing countries to use sustainable agriculture for economic development and food security. This chapter outlines the problem of food security and the opportunity that it presents.

---

[1] This chapter is based on work published previously: 'Polimeni, J. M., Iorgulescu, R. I., Bălan, M. (2013). Food Safety, Food Security and Environmental Risks. *Internal Auditing & Risk Management*, 1(29): 53-68'

2.1. INTRODUCTION

At the advent of the nineteenth century, Thomas Malthus envisioned a future where population would increase faster than food production. As a result, he predicted the specter of mass death through starvation. To prevent this prediction from becoming a widespread reality, the approach, until now, has been to convert more land for agricultural purposes and use technological advancement s, such as fertilizers and GMOs, new and more sophisticated farming techniques, and capital investment. We say 'widespread' because even though the supply of food produced today could feed the whole world, millions of people go to bed hungry or even starve to death (Badgley et al., 2007). The explanation lies with global agricultural policies, the inadequate structure of food production and distribution systems, and with the increasing global income inequality and the terribly wasteful 'modern' lifestyle. In our 'global village', very often the poor cannot afford to purchase or even produce food and in extreme cases they have to rely on 'humanitarian aid.' To make things worse, climate change increasingly affects the delicate balance needed for a stable (as much as possible) production of food. As a result, given the projections for world population in the next twenty-five years, it becomes more and more uncertain if the amount of food currently produced globally can meet the needs of these additional people. Therefore, food security is a real issue that countries and regions must confront.

Food security is defined here as the ability to obtain an adequate supply of quality food that is safe and culturally acceptable. Figure 2.1 illustrates, in a simplistic way, the determinants of food security at the household level.

A household (urban or rural) could obtain their food either through its own production or by buying it from a seller (direct producer or intermediary) as raw or processed food. In both situations there are agriculture-related factors and economic factors influencing the household's ability to afford and obtain the necessary food.

In the past few decades another trend inextricably linked to economic growth increasingly rang the food security alarm: large-scale urbanization projected to grow in the coming decades with the swelling population. Urbanization causes several food security issues. First, urbanization removes labor force from agriculture as individuals move to urban centers for higher incomes. With a reduction in labor there is additional stress on the ability to produce agricultural products. Second, and related to the first, is that urbanization typically leads to higher incomes for those individuals who migrate enabling households to purchase and consume greater amounts of food.

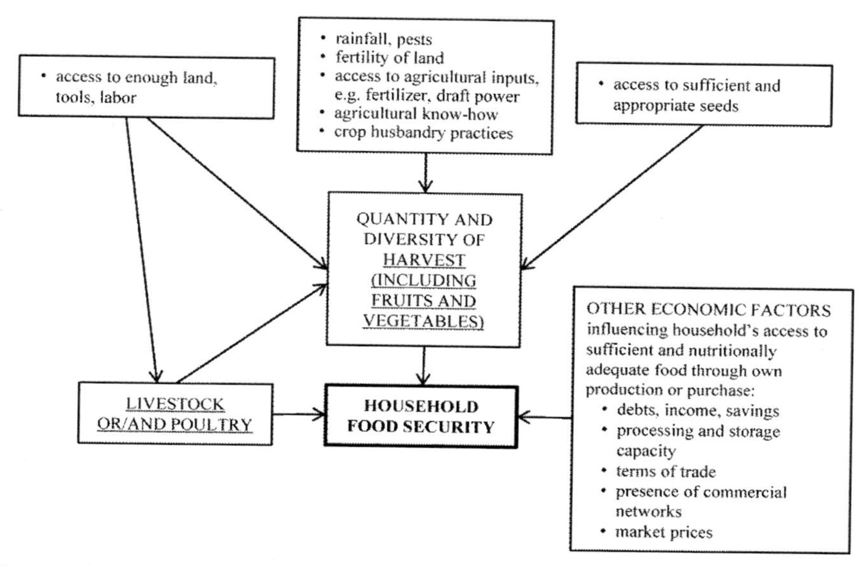

Source: Authors' adaptation after FAO (1995).

Figure 2.1. Determinants of Household Food Security.

According to Engels' Law, with increasing income people demand more and better food, they just spend a smaller percentage of their total income on food (Pritchett and Spivack, 2013). As a result, with the increase of the 'richer' urban population the demand for food rises, as does the pressure for additional agricultural production. Alternatively, the increased demand for food could be solved through imports. Obviously, importing food is a form of 'vital' dependence, but it decreases food security. If, due to various circumstances in the exporting country, food becomes scarce the social repercussions in the importing country could be dramatic as they will no longer be able to obtain food. The following pages present how sustainable agriculture can increase food security as well as be used as a tool for economic prosperity.

2.2. The Connection between Oil and Food Prices

Despite the fact that developed countries became food secure and successfully completed a structural transformation of their economies, many, if not all, are arguably food insecure today, relying on industrial agricultural approaches which have led to monocropping and importing food. For example, even though the United States is among the largest agricultural producing countries, much of their food that is consumed is imported. Any food crisis – whether due to shortage or to a spike in prices – can cause political and economic instability. This instability forces households to spend more of their income on food and increases their precautionary savings for periods of uncertainty. In turn, these changes by households can have a significant spillover effect on the rest of the economy. Those looking to profit turn to speculation, rather than productive investment, slowing economic growth (Timmer, 1998b: 207).

For example, food prices, particularly cereals, spiked in 2008 contributing to the financial crisis that occurred late that year. Once again, food prices spiked in 2010 and 2011 (Figure 2.2) causing many forecasters to speculate that further economic, political, and social disruption will follow. In fact, one news report suggested that the end of cheap food may be drawing to a close (Arasu, 2011).

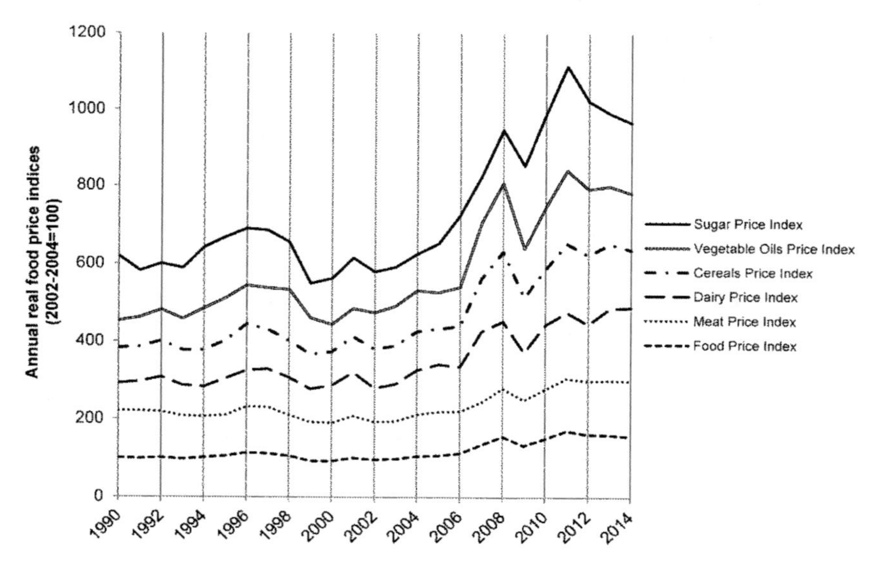

Source: Authors' processing of FAO data (Note: All indices have been deflated using the World Bank Manufactures Unit Value Index (MUV) rebased from 2005=100 to 2002-2004=100).

Figure 2.2. Stacked lines chart of Annual Real Food Price Indices (2002–2004 = 100).

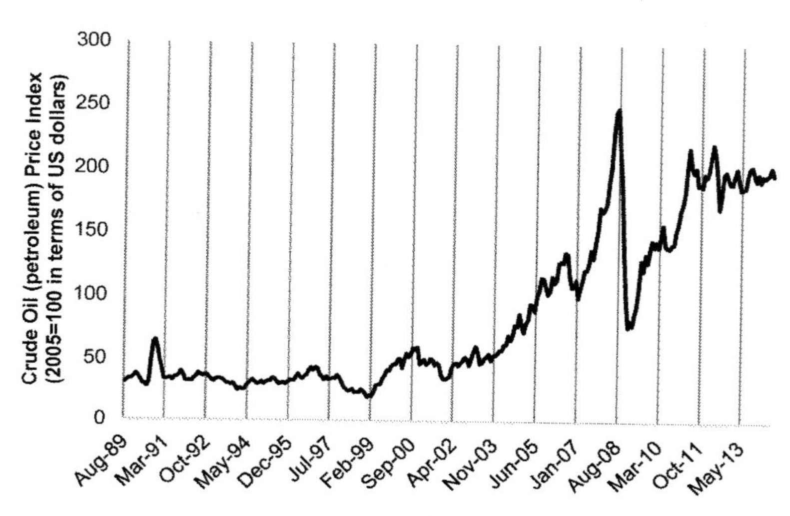

Source: Authors' processing of IMF data (Average Petroleum Spot Price: equally weighted average of three spot prices: Dated Brent, West Texas Intermediate, and the Dubai Fateh).

Figure 2.3. Crude Oil (petroleum) Price Index (2005=100 in terms of US dollars).

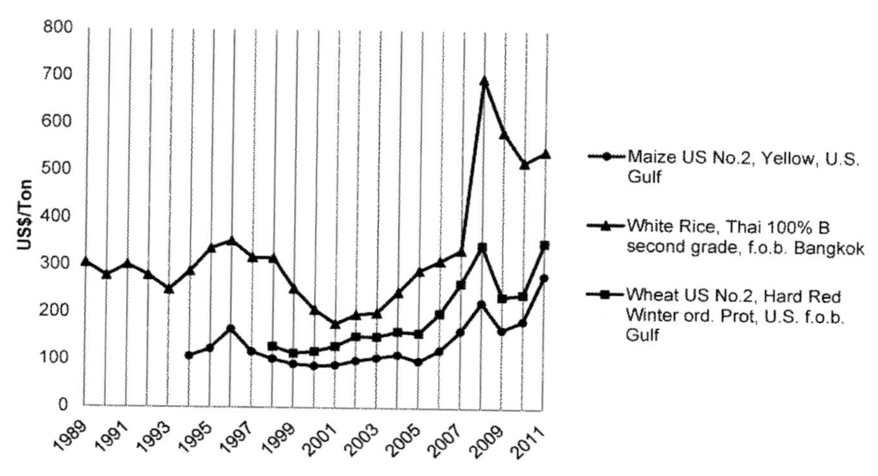

Source: Authors' processing of FAO data.

Figure 2.4. Average Annual Prices for Key Agricultural Commodities (Maize, Rice, and Wheat).

As Figure 2.2 illustrates, commodity food prices have risen significantly since the middle of 2010. These prices are expected to continue to rise as energy prices increase (Figure 2.2 and 2.3) due to the unrest in the Middle East (i.e., The Arab Spring). In fact, financial and economic analysts projected the price reaching the $10 per metric ton mark in 2011 because of oil price spikes.

Figure 2.4 illustrates the average annual prices for the key agricultural products of wheat, rice, and maize.

According to Fuglie, MacDonald, and Ball (2007) the evolution over time of the agricultural output price index and the agricultural input price index is correlated with oil prices. As expected for an industrial agriculture heavily dependent on products derived from the petrochemical industry, Figure 2.4 shows that food prices have a strong variability linked to petroleum prices and 'feel' the same shocks. This price variability has rendered bankrupt those small farms that were not able to adapt.

While the intervals for the variation of commodity food prices are uncertain, one thing is sure – food prices will remain at a higher level and will be more volatile than the world has experienced in the past forty to fifty years. As many countries have little agricultural production and/or infrastructure in place, the increase in food prices is highly likely to cause food security issues. Consequently, a top priority should be implementing and/or sustaining a diversified and sustainable food production system to create food security.

This form of agriculture will counter the speculators investing in commodities and reduce their effects on food prices. Furthermore, focusing on food security through sustainable agriculture would ensure that a nation would not be dependent on food imports. Economic growth could follow as food prices would be stable, domestic farmers would have greater security as their products will have a market, and rural households would have more disposable income to spend spurring rural economic development.

2.3. Food Dependency, Economic Downturns, and the Role of Local Food Production

The potential effects of an insufficient and/or nutritionally inadequate supply of food range from health status to labor implications. Therefore, any disruption of the food supply can quickly cause panic among a population.

Food access became more problematic when the global economic crisis started in 2008. The economic downturn, largely caused by higher energy prices and a breakdown in financial institutions, has led to higher food and commodity prices, a decrease in profits from exports, and less income that could be used for food purchases or for remittances. The importance of food security, from a political and scientific point of view, was emphasized when higher food prices caused riots in more than two dozen countries (Barrett, 2010); in many cases, the riots lasted more than five months and caused a string of bankruptcies (Singh, 2011).

In early 2011, the Middle East and Northern Africa experienced a new instance of food price inflation generating more social upheaval. The protests had pushed the overthrow of the governments in Tunisia and Egypt. Moreover, at that time, Jordan, Lebanon, and Algeria were on the verge of complete turmoil, while in Syria and Libya the protests became civil war s; unfortunately some of those countries are still in a state of chaos (mostly Syria and increasingly spilling over to Lebanon and Iraq). In this case, the increase in food prices was the main cause for social unrest in developing countries with little agricultural production and high youth unemployment.

Agriculture is a vital sector for fueling economic growth. Countries, developed and developing alike, rely, to varying degrees, on food imports. As a result, agricultural exports have become big business, especially for developed countries. However, economic downturns reduce countries' access to imported food and allows for the emergence of food insecurity. Rosen and Shapouri (2009: 40) show that over three decades, from 1970 to 2003, the least developed countries had the highest increase of dependence on food imports.

In 2003, in low income countries, imports accounted for 17% of grain consumption, 45% for sugar and sweeteners, and 55% for vegetable oil s, an increase compared to 1970 levels from 8%, 18%, and 9% respectively (Rosen and Shapouri, 2009: 39). Since developing countries have the weakest influence on world market prices, they experience the greatest negative impact of globalization and international trade. Environmental problems, limits of arable land, water constraints, and an increasing reliance on agricultural products for energy become another threat to food security through increasing agricultural commodity prices.

A long-term solution to food security issues would be to increase domestic agricultural productivity without increasing environmental degradation and concurrently addressing climate change issues. One way to reduce vulnerability to food insecurity is to grow a variety of agricultural products using a sustainable agriculture approach. At the same time this approach would encourage economic growth.

In order to use sustainable agriculture as an economic development tool, it is necessary to understand why agriculture, a primary production activity in most developing countries, has not been successful. In general, agriculture has performed poorly in developing countries because it has been neglected while governments adopted public-policies largely concentrated on the development of industry.

However, the majority of rural families in developing countries are engaged in agricultural production, not only as a source of income but also as a source of food, clothing, and shelter. Furthermore, agriculture is a way of life for people farming the land. Since there are so many people farming in relation to the available land, labor is the primary input to production while technology, business organization, and physical capital are limited in use. So, any change to farming methods is a change in the farmer's way of life and transforming the agricultural system means changing the social and political institutions within countries. Reducing the institutional obstacles is necessary for rural and agricultural development. Therefore, obtaining maximum levels of agricultural output for the country is difficult.

2.4. Impacts of Industrial Agriculture

Industrial Western-style agriculture with increased output yields and decreased costs is synonymous with the Green Revolution. The revolutionary 'recipe' combines the intensive use of irrigation, machinery, improved biological varieties, synthetic fertilizers and pesticides with modern management techniques. The next subsections will discuss three impacts of this agricultural approach, namely farm concentration and specialization, the reduction in the number of farms and negative environmental effects.

2.4.1. Farm Concentration and Specialization

Monocropping or high-yielding crop varieties grown in industrial fashion is one way to intensify agricultural output. The high profits obtained when increasing the size of the plot of land under monocropping cultivation motivates industrial agricultural firms to use more expensive inputs (e.g., chemical fertilizers), increase production, and thus lower their per unit cost. At a country level, this practice is encouraged through agricultural policies.

The increasing global trend towards importation of agricultural products steered agricultural practices towards monocropping with this approach becoming prevalent in modern agriculture. Additionally, increasing food imports amplified competition pushing some domestic farmers to quit farming and encouraging 'land grabbing'. It should be also noted that farmers in poor countries usually cannot compete with subsidized farmers from rich countries. Unsurprisingly, the process of monocropping expansion leads to farm concentration and a reduction in the variety of crops produced.

In the case of the United States, Figure 2.5 illustrates the decrease in the number of commodities produced per farm over the past century.

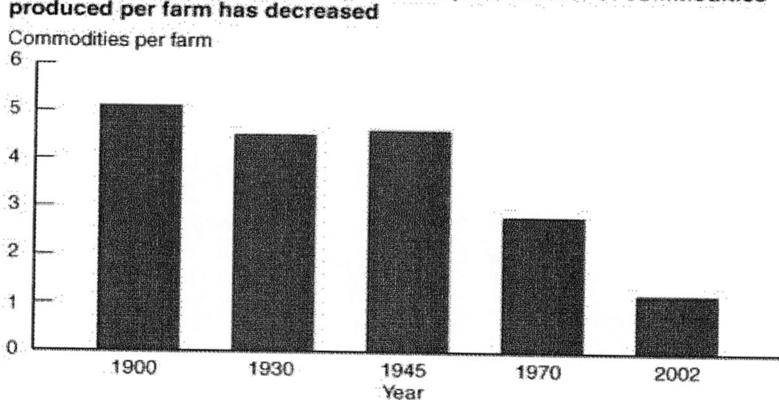

Note: The average number of commodities per farm is a simple average of the number of farms producing different commodities (corn, sorghum, wheat, oats, barley, rice, soybeans, peanuts, alfalfa, cotton, tobacco, sugar beets, potatoes, cattle, pigs, sheep, and chickens) divided by the total number of farms.
Source: Compiled by Economic Research Service, USDA, using data from *Census of Agriculture, Census of the United States*, and Gardner (2002).

Source: Dimitri, Effland, and Conklin (2005) p. 5.

Figure 2.5. Increasing Farm Specialization in the United States: 1900-2002.

For example, on almost 70% of the agricultural land in the Midwestern United States only corn, soy, and sugar are grown; the farms have an average surface of 14,000 acres (Barber, 2005). Another example is California which generates half of the tomatoes produced in the U.S., more than two-thirds of strawberries and lettuce, almost all the grapes (approximately 93%) and all the almonds. The concentration goes even deeper: more than half of Californian grapes come from three neighboring counties and more than two-thirds of the lettuce from six neighboring counties (Cameron and Pate, 2001).

Similarly to the U.S. case, farm specialization and the decrease of the number of farm commodities produced per farm was reported in other Western countries that allowed and even encouraged the industrial agricultural approach.

2.4.2. Worldwide Decrease in the Number of Farms

Unfortunately, the situation got even worse since the trend towards an increase of the farm size is often accompanied by a decrease in the number of farms in developed countries where most of the production occurs.

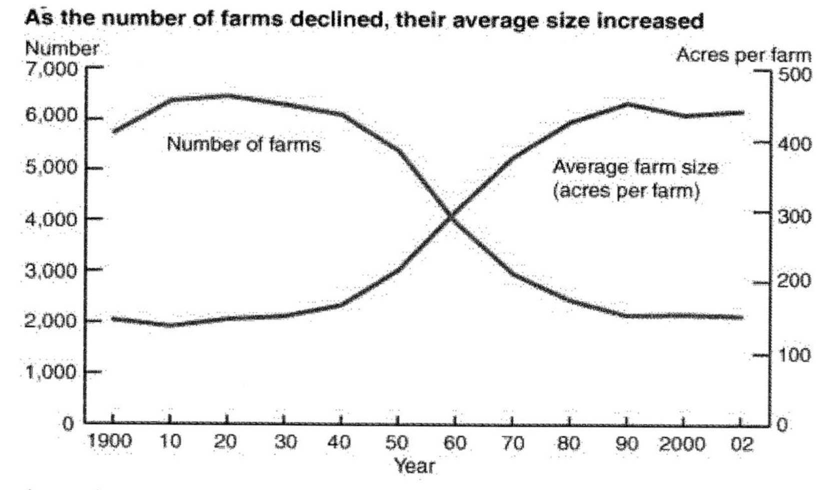

Source: Dimitri, C., Effland, A., and Conklin, N. (2005) p. 5.

Figure 2.6. Average Farm Size vs. Number of Farms in the U.S.: 1900-2002.

Figure 2.6 shows, for the case of the United States, this inverse relationship between average farm size and the number of farms. Other Western countries have a similar relationship. In large part these changes to the structure of farms have come as a result of globalization. As Western countries have increasingly shifted their production to grow just a few crops for biofuels and other non-food purposes, developing countries are exporting their agricultural production for money, taking food out of their system and, on many occasions, making their countries food insecure.

For example, India has been successful in increasing agricultural production the past decade or so, yet the country still has a high percentage of citizens which are either under- or malnourished because they export their food crops. Although some may think this to be an isolated case, due to financial considerations many developing countries are also exporting their agricultural production. The result has been for small farms in developed countries and developing countries alike, to go out of business while large farms relying on the industrial agricultural method of monocropping has become the norm. Many of the small farms that remain are so small that they can only be of the subsistence or semi-subsistence variety (Lowder et al., 2014). Governments, using neoclassical economic theory, have promoted industrial agriculture with mega-sized farms to shift labor to non-agricultural purposes because technological advancement s. As discussed previously, this would cause the prices of agricultural products to decrease. Furthermore, neoclassical theory also encourages that resources be employed in non-agricultural sectors because the returns will be greater in these other sectors now that agricultural prices are lower.

2.4.3. Environmental Impact of Industrial Agriculture

The high profits of monocropping come at the great cost of environmental damage. This statement is particularly true for communities or countries which diminish, destroy, or use-up their natural resources that are needed to produce food, as well as for their survival.

Intensive monocropping depletes nutrients from the soil causing fertility to decrease and forcing farmers to apply even higher levels of chemical fertilizers and pesticides to produce the same amount of agricultural output. Unfortunately, there is a turning point beyond which yields start decreasing and in the end the farmland becomes brown-field or land not suitable for agricultural use. Fuglie, MacDonald, and Ball (2007, Figure 1) illustrate this phenomenon using total farm output, total inputs, and total farm productivity in the United States from 1948 to 2004. After 1998, total farm output in the United States has flattened out.

Unfortunately, until recently, environmental services, such as the absorption of the residuals from agricultural production, have been treated as a free good. This point of view allowed for scientific and technical innovation, such as man-made chemicals and management systems, to be treated as perfect substitutes for land. Overvaluation of land and undervaluation of the social costs of the negative externalities of the industrial agricultural process only followed as logical results (Runge et al., 1990; Ruttan, 1994).

In addition to pollution caused by fertilizer use, agricultural land has also been negatively affected by other environmental problems such as polluted water and acid rain. Many of these issues came to the forefront during the late 1980s and the early 1990s with the release of the Brundtland report and the UN conference on the environment and development in Rio de Janeiro (Staatz and Eicher, 1998). Later, in the late 1990s and early 2000s climate change became an important issue with the global populace and concern grew over how agricultural production would be affected. The increased levels of man-made gases in the atmosphere (such as the higher levels of carbon dioxide due to massive deforestation) as well as other human caused environmental problems are believed to be a major contributor to climate change; agricultural systems will be, certainly, negatively impacted. Simulations using various models have been used to illustrate how global agricultural systems may be affected (Cline, 2007; Mendelsohn and Dinar, 2009; Lobell and Burke, 2010). Not surprisingly, developing countries will be the biggest losers.

Modern industrial agriculture is caught in a vicious feedback loop: increasing amounts of synthetic chemicals are used to increase production and as a result, in time, the agricultural land is degraded, groundwater becomes polluted and also the embedding ecosystem. In turn, the contaminated water and animal excrement are returned to the land which further deteriorates the land and negatively impact output yields.

2.5. Stages in Economic and Agricultural Development

We are still left with the question how to increase agricultural output despite several constraints. First, land is a fixed resource. Agriculture competes with the residential, commercial, and industrial sectors for productive land. Since these other sectors are valued by decision-makers more highly then the agricultural sector, land used for farming has slowly, over time, diminished. Nearly all of the world's prime agricultural land is either already being farmed or has been converted for other uses. Urban migration, a global phenomenon, has only exacerbated this land conversion. This is a particularly troubling situation in developing countries because, unless technology changes, the labor that is needed to produce agricultural goods decreases over time limiting the amount of output that can be used to feed growing populations.

Understanding the stages of agricultural development is necessary to design an economic development policy focused on agriculture. Eicher and Staatz (1998: 71) outline four stages of development, which are summarized below.

The first stage is typified by most of the labor force in a country working in agriculture. Primarily, this type of agriculture would be for subsistence purposes and is common in many developing countries. For agriculture to contribute to economic development at this stage, significant investment in infrastructure and research must occur.

Once the second stage is reached, agriculture begins to make contributions to economic growth, both in the agricultural sector as well as in other economic sectors. At this stage, growth comes largely from improvements in technology, as well as improved markets, institutions, and agricultural policies that enhance agricultural production.

By stage three, the agricultural sector is integrated with the rest of the economy through more mature and efficient markets that connect the rural and urban economies.

By the fourth stage, the total labor force in agriculture falls to below 20%, subsidies are commonly used to maintain a farm production structure and encourage overproduction to keep food prices low, and the percentage of household expenditures on food decreases to less than 30% of total income. For example, in 2012 Americans spent 6.6% of their household income on food (Wolfe, 2014) while in a country such as Romania households spent 42% of their income on food (Lever, 2013).

Interestingly enough, as the agricultural sector progresses through each of these stages its share in GDP decreases. According to the World Bank (1982), agricultural growth for many countries generally lags behind if annual GDP growth is 3% or higher. This finding is consistent with the four stages of agricultural development. For stages one and two agricultural production can only increase with technological change and improved markets. For instance, the creation or improvement of transportation systems, improved organic farming techniques, and improved commercial markets, all are examples of technological and market improvements that can expand agricultural production. However, one must keep in mind that rural households and the agricultural products that they sell are the basic market for a large variety of consumer goods that spur economic development in other economic sectors. Therefore, agricultural development must be part of any national economic development program.

As economies develop over time the importance of agriculture usually declines. This paradox helped to form the neoclassical economic viewpoint that agriculture is a declining sector that is nothing more than a support system to provide labor and food to industry and other economic sectors (Timmer 1998a: 122). Additionally, in the neoclassical economic viewpoint agricultural productivity enables rural dwellers to purchase industrial and commercial products made in urban centers, and any surplus agricultural products would be exported to acquire imports that will spur further economic growth. These ideas can be traced back to the works of Clark (1940) and Kuznets (1966). According to the ideas presented by Clark, Kuznets and many other neoclassical development economists, the agricultural sector is the support system for the rest of the economy, providing key resources so that the economy can be transformed into an industrial, commercial, and service economy. For this objective to be accomplished public policy must be designed to encourage these industries, creating a structural transformation of an economy. However, no country has ever achieved rapid economic expansion without first being food secure (Timmer 1998b: 205).

2.6. Sustainable Agriculture, Integrated Rural Development and Poverty Reduction

Sustainable agricultural growth can significantly reduce poverty. First, as the sale of agricultural products increases, the demand for labor will increase also in rural regions in both the agricultural and non-agricultural sector s. Agricultural industries will require products of support industries and the corresponding increase in wages will increase household disposable income, increasing the demand for non-agricultural products. Second, agricultural growth will lead to stable prices for agricultural products which is a price reduction in real terms under normal inflationary conditions. Both of these two points will result in an increase in real disposable income for both rural and urban households, increasing the demand for both agricultural and non-agricultural products. In turn, the demand for labor will rapidly increase as the economy expands. Therefore, the rural-urban dynamic leads to economic development for the entire economy.

While this economic development scenario sounds like a plausible and easy scenario to carry-out, the reality is that few, if any, countries develop agricultural policy in this way. A number of reasons exist for the failure to implement strategies, such as those listed above, that will reduce rural poverty. The agricultural subsidies that are given by governments to farmers are the primary reason. These subsidies, particularly those provided by developed countries, encourage the over-production of agricultural products which lowers the world price for food. Since food prices are artificially kept low, farmers in developing countries cannot afford to commercially produce food and agricultural products. Therefore, agricultural production in developing countries largely falls under the subsistence or semi-subsistence category. The result is high unemployment in rural regions causing people to flock to urban centers seeking employment and higher wages. Furthermore, as rural people migrate to cities, the urban centers are under increasing stress from the additional population causing urban unemployment and problems.

Although agricultural subsidies are the main reason agricultural policies developed for rural economic development are not implemented, a variety of other reasons exist for an unsuccessful integrated rural development program. First, and related to agricultural subsidies, there is a lack of commitment from individual and collective governmental agencies and politicians (World Bank, 1987). This statement is true for both developed and developing countries. Governments and politicians in developed countries do not want to alter agricultural policy because the agricultural corporations and farmers in their respective countries have considerable influence through lobbying groups and political contributions. In developed countries the agricultural corporations and large-scale farmers benefit from a lack of a cohesive rural development strategy. Therefore, the politicians and government agencies are often afraid to challenge the agricultural corporations and farmers because they are fearful of losing political power as well as monetary contributions from these special interest groups. This lack of government commitment and power of the agricultural industries and large-scale farmers create an adverse policy environment that hampers agricultural policy reform from occurring.

A second reason for an unsuccessful integrated rural development program is the lack of infrastructure. This reason is more generally a problem for developing countries as infrastructure is often limited due to insufficient funds to carry-out expensive development projects. In particular, the lack of revenue has a major impact on the accessibility of technology (World Bank,

1987). Infrastructure, such as roads, food processing plants, and railways are needed to expand rural development. Without proper infrastructure additional agricultural production will not occur because farmers will not grow additional product that they cannot bring to market to sell. Therefore, infrastructure is a necessary investment in rural areas if a country seeks to increase agricultural development. Additionally, initial investment in technological improvements such as irrigation systems and tractors would most likely need to be financed by national governments until the agricultural system moved from subsistence or semi-subsistence agriculture, stage one, to either stage two or higher in agricultural development. However, these issues also exist because in many countries struggling to achieve development there is a lack of institutions in rural regions. To facilitate rural development local and regional institutions, such as agriculture agencies, are needed to monitor development programs (World Bank, 1987).

However, perhaps most important is access to financial services for rural dwellers. Often in rural areas there is either a severe shortage of or no financial institutions at all for people living in these regions to receive credit or to save their earnings. While in recent years microfinancing has filled this void somewhat, there is still a large problem for those living in rural regions to obtain financing. Access to financial services is a vital component of rural economic development because without access to these services farmers cannot make the appropriate investments they need to expand their productive capacity.

A third reason for an unsuccessful integrated rural development program is that the benefits for participating in agricultural development are often unrecognized. Farmers, especially in developing countries, do not see the point in expanding their production if the support programs and institutions, such as those listed above, are not available because they do not have access to markets that are large enough to sell their goods. From a governmental perspective, financing of these initiatives must occur without a guarantee that agricultural development will occur. The government must convince farmers that if they produce more they will earn more money and be better off. In many countries this obstacle is very difficult to overcome as many citizens do not trust their government because of bad previous experiences in their relations. Therefore, a major problem is the lack of beneficiary participation (World Bank, 1987).

Old and new research shows that small farms can be more productive and sustainable than larger farms (Berry and Cline, 1979; Maass Wolfenson, 2013). Fortunately, history provides a number of examples to examine so we do not have to rely solely on conjecture of anecdotal information. One of the first examples dates back to the Middle Ages when integrated crop-animal husbandry replaced the two and three-field systems. The husbandry system introduced and used intensively forage and green manure crops to maintain and improve soil fertility (van Bath, 1963; Boserup, 1965). Another example comes from East Asia. Animal and human manures were recycled along with grain and straw in an area where the nutrients and minerals could be carried to the fields with the irrigation water (Hayami and Ruttan, 1985).

These examples provide evidence that sustainable agriculture is an acceptable alternative to modern, industrial agricultural systems. However, sustainable agriculture has not gained a considerable amount of market share for a number of reasons.

First, despite increasing evidence that sustainable agriculture can produce food at levels competitive with industrial agriculture, many nations and individuals have major concerns that the amount of food produced using a sustainable approach would not be substantial enough to feed the current population of the world, nor the projected population in the future. In the past the solution to increase agricultural production was simple; to increase the amount of land used in production.

However, over time the most fertile land was already in use and agriculture was in an ever-increasing competition for land with industrial, commercial, and residential developers. Therefore, the land suitable for agriculture is already in some form of use, leaving only marginal land, at best, for farming.

This realization has led to the second concern nations and individuals have about sustainable agriculture which is related to producing sufficient output yields. In the industrial agriculture approach farmers have turned to fertilizer, pesticides (including herbicides), and irrigation to increase output yields. However, sustainable agriculture does not use artificial, external inputs to increase output yields.

This reality has caused concern that productivity growth using sustainable agriculture will not be possible. Therefore, reliance on industrial agriculture has continued despite increased calls and demand for sustainable agricultural approaches.

2.7. The Positive Impact of Local, Sustainable Agricultural Practices

Thus, neoclassical economic theory has contributed to the agricultural conundrum that exists today; large, mega-farms producing relatively few crops. However, these same policies in conjunction with environmental concerns and apprehension over population pressures has caused many individuals to understand that the current industrial agricultural methods will result in lower agricultural productivity, higher food prices, and increased poverty in rural regions (Eicher and Staatz, 1998). This realization has resulted in a demand driven movement for sustainable agricultural products that has significantly expanded in the past decade.

Using sustainable agricultural techniques can halt, and with some methods reverse, the negative externalities caused by industrial agriculture. Rural household incomes and living standards, given some external assistance such as those listed previously, can increase substantially using a sustainable approach to agriculture.

Agricultural products obtained using sustainable methods have higher value-added and cut costs due to the reduction of external inputs to production and the goods are sold locally. However, there are indirect benefits to sustainable agriculture as well. Rural economic development will lead to improved public health, better public services such as education and sewage treatment, a cleaner environment, and improved rural-urban equity.

Although sustainable agricultural practices can have a positive impact on the development of an economy alone, to achieve the greatest benefit these practices should be complemented by other development policies. Improved macroeconomic and microeconomic policies, in conjunction with emphasizing sustainable agriculture, can stimulate and stabilize an economy and improve food security more quickly. For example, for developing countries where typically 40% to 50% of GDP, 70% to 80% of the labor force, and 70% to 90% of earnings from foreign exchange comes from agriculture, economic growth is nearly impossible to sustain without expanding the rural economy (Timmer, 1998b: 200).

Furthermore, sustainable agricultural practices can also alleviate food inflation risks. Since food is not imported from long distances, transaction costs are not incurred because food is available locally. Additionally, if food was produced and sold locally, world prices would drop keeping food inflation in check.

Therefore, sustainable agriculture can substantially reduce poverty in a country. The most direct and obvious impact will be the reduction on rural poverty. Sustainable agriculture will increase employment in rural regions as more farms means that more workforce is needed. This increase in employment, in turn, leads to more disposable income creating more prospects for service companies and self-employment opportunities to meet the new, increased demand from rural households. As a result rural wages increase which puts less pressure on urban systems because rural migration will substantially lessen. In turn cities and countries do not have to spend as much money to expand and improve urban infrastructure and these savings can now be spread to other parts of the economy.

According to the United States FDA (Food & Water Watch, 2008) imported fruit is four times more likely and vegetables twice as likely to have illegal levels of pesticide residues. Other Western countries are likely to have similar results. Beru and Salisbury (2002) reported that imported produce to the United States was more than three times more likely to contain *Salmonella* and *Shigella* than domestic produce.

Due to various international trade agreements such as the World Trade Organization, the European Commission, and the North American Free Trade Agreement the levels of food actively traded has reached unprecedented heights which has put undue pressure on inspection centers and reduced food safety.

As shown in Figure 2.7 import shares of U.S. food consumption has steadily increased from 1981 to 2001 in every major food consumption category. From 1990 to 2007, total fresh and processed fruit and vegetable imports in the United States have more than doubled (USDA).

According to the FAO, food imports in developing countries are projected to increase considerably in the coming years in most major food consumption categories.

Perhaps the greatest indirect impact on economic development from using local, sustainable agricultural practices is the reduction in health care costs, both monetarily and with improved health of workers which leads to higher productivity. There are a few reasons for health improvements. First, because food is grown locally and does not have to be picked before maturity in order to be shipped thousands of miles, the food will be more nutritious. Second, agricultural products produced using sustainable methods do not have harmful pesticide residues.

Summary import shares of U.S. food consumption[1]

	Average 1981-85	Average 1986-90	Average 1991-95	1996	1997	1998	1999	2000	2001	Average 1997-2001
					Percent					
Average import share of total food consumed	7.8	8.6	9.4	10.4	10.9	11.2	11.2	11.1	11.2	11.1
Animal products[2]	3.4	3.7	3.5	3.5	3.6	4.4	4.5	4.5	4.6	4.3
Red meat	6.7	8.1	7.3	6.4	7.1	7.7	8.2	8.9	9.3	8.2
Dairy products	1.9	1.8	1.9	2.0	1.9	2.9	2.9	2.7	2.8	2.6
Fish and shellfish	50.9	56.0	56.0	58.5	62.1	64.7	68.1	68.3	68.2	66.3
Animal fat	1.0	1.6	3.5	3.6	5.9	5.5	5.9	6.9	2.6	5.4
Crops and products[3]	11.5	12.6	13.9	15.6	16.3	16.3	16.3	16.2	16.4	16.3
Fruits, juices and nuts	14.1	17.9	17.0	16.9	18.8	18.4	21.0	21.9	21.4	20.3
Vegetables	4.7	6.0	5.9	7.8	7.9	9.0	9.0	8.7	9.4	8.8
Vegetable oils	15.5	17.6	17.4	19.3	19.4	16.3	16.9	18.3	15.5	17.3
Grain products	1.1	2.3	4.8	3.9	5.8	5.3	5.3	4.6	4.9	5.2
Sweeteners and candy	21.6	11.9	10.8	15.2	16.1	12.9	11.0	9.8	9.7	11.9

[1]Calculated from units of weight, weight equivalents, or content weight.
[2]Import shares of poultry and eggs are negligible, but accounted for. Red meats are estimated from carcass weights.
[3]Includes coffee, cocoa, and tea whose import shares are 100 percent; also includes wine and beer.

Source: Jerardo (2003), p.2.

Figure 2.7. Import Shares of U.S. Food Consumption 1981-2001.

2.8. Food Scares and the Threat of Agricultural Bioterrorism

Domestic agricultural products, especially in developed countries, are much less likely to have harmful bacteria. That does not mean that foodborne illnesses do not occur domestically. For example, the United States has had a number of high profile outbreaks of *Salmonella* in the past few years. Additionally, the United Kingdom experienced a large outbreak of foot-and-mouth disease in 2001. However, domestically produced food will typically be safer to consume. First, domestically produced agricultural goods are easier to inspect, whereas the amount of imported food is very large and requires an extensive inspection system. Secondly, consumers have a resistance to the bacteria in domestically produced food while imported food will most certainly contain bacteria that are foreign to their bodies leaving them susceptible to disease and illness. Lastly, if an outbreak does occur, domestically produced agricultural products are much easier to track in order to find the source of the outbreak. Imported agricultural products are combined with domestic products which severely complicates the ability to find a contamination source because the food has been cross-contaminated.

The food security issues associated with imports has brought the problem of agricultural bioterrorism to the forefront. Take, for example, a statement made in December 2004 by then Secretary of Health and Human Services Tommy Thompson, "I, for the life of me, cannot understand why the terrorists have not attacked our food supply, because it is so easy to do so." (Boyle, 2005) Additionally, the World Health Organization has warned, "... the malicious contamination of food for terrorist purposes is a real and current threat" (WHO, 2002). Dr. Donald Henderson, an expert on biological terrorism and the former director of the U.S. Office of Public Health Preparedness stated, "At least ten countries are now engaged in developing and producing biological weapons ... these will eventually be misused." (Henderson, 2001) Pre-harvest crops are especially vulnerable because any disruption in the growth of crops will cause existing food stocks to be used quickly, leading to a food shortage. The centralized agricultural system that currently dominates the Western approach to agriculture is a weakness and makes agriculture an attractive target.

Pathogens that can be used for agricultural destruction are much more easily acquired and disseminated than those pathogens intended to do harm on humans. Furthermore, some agricultural produce generates toxins that can create severe sickness or death when the crops are contaminated with specific pathogens. On many occasions diseases to agricultural crops have caused unintentional contamination. For example, the World Health Organization estimates that, annually, 33% of people in industrialized countries suffer from disease caused by contaminated food (FDA, 2003; WHO and WTO, 2002). Moreover, the Centers for Disease Control estimate that nearly 25% of Americans will suffer from accidental food contaminated by pathogens every year (FDA, 2003; U.S. Census Bureau, 2001). Imagine the destructive effects and resulting consequences of a deliberate terrorist attack contaminating the food supply. The First Annual Report to the President and The Congress of the Advisory Panel to Assess Domestic Response Capabilities for Terrorism Involving Weapons of Mass Destruction published in 1999 states that "... a successful attack could result in local or regional destabilization" and have a large impact on international commerce since so much of the industrial agricultural approach depends upon global trade.

While the costs of an attack could potentially be trillions of dollars, the biggest cost could be psychological as consumer confidence is considerably weakened. The resulting chaos would make the social disruption from food price spikes mentioned earlier in the chapter seem mild in comparison. The panic would spread further as the international food trade is disrupted and then halted as countries attempt to stock-up and secure their food supply. World Trade Organization members are required to prohibit imports of plant or animal products that could introduce diseases into their countries, further increasing the likelihood of a food shortage (Wheelis et al., 2002; Wheelis, 1999; FAS, 2001). Every country would feel the global economic impact.

Probably the most plausible and cost-effective counter-terrorist measure would be to encourage more diverse, independent, sustainable agricultural entities, in particular the promotion of farms more dependent upon local sources of resource inputs and catering to local markets. Sustainable agricultural approaches can significantly reduce the probability of an agro-terrorist event because of the diverse selection of produce that make the complete destruction of its crops unlikely. Additionally, sustainable farms are much smaller and localized than the industrial farms that are commonplace today.

CONCLUSION

The reliance on industrial agriculture has limited the capacity of countries to respond to the concerns of food shortages, particularly in developing countries where there is considerable difficulty in developing and maintaining agricultural research (Eicher, 1994). Furthermore, the external inputs used in industrial agricultural practices, fertilizer, pesticides, and irrigation, are heavily energy intensive. As stated previously, ever-increasing amounts of these external inputs will be needed to maintain production yields consuming more energy. However, as energy prices increase these external inputs will increase the costs of production in the industrial agricultural approach and become a significant primary resource constraint to expanding production further (Desai and Gandhi, 1990; Chapman and Barker, 1991).

Despite all the advantages of a sustainable agricultural approach for economic development, the environment, and society, industrial agriculture remains the leading method today. Countries and policy-makers around the world have been led to believe that agricultural commodities can be treated like any other product and traded on the global marketplace. This belief has become the dominant viewpoint based upon neoclassical economic theory and the concept of comparative advantage which stresses that each country has their own development path given resource endowments and their stage of development. The result was economic policies promoted by economic development agencies that stressed industrialization for developing countries which resulted in monetary resources being diverted from agriculture. Unfortunately, in many instances, those economic policies led to stagnant economic growth and countries that were once food secure are now relying on imports of agricultural products. In developed countries the impact has been slightly less severe. In these countries, as development expanded, both population and household incomes rose, which increased the demand for food. To feed this economic growth, labor was shifted from the agricultural sector to nonagricultural production. The key difference between developed and developing countries is that developed countries already have high agricultural production and the necessary support infrastructure. However, whether a developing country or a developed country, it is clear that agricultural development is a real opportunity to raise national income and improve the welfare of rural dwellers. This chapter has shown this idea to be valid. Furthermore, this chapter has presented a clearly stated case as to why sustainable agriculture is the best method for agricultural development with the largest opportunity for economic development.

References

Arasu, K.T. (2011). An Era of Cheap Food May Be Drawing to a Close. *Reuters News Service*. http://www.msnbc.msn.com/cleanprint/CleanPrint Proxy.aspx?1296437095005

Badgley, C., Moghtader, J., Quintero, E., Zakem, E., Chappell, M. J., Aviles-Vazquez, K., Samulon, A., & Perfecto, I. (2007). Organic agriculture and the global food supply. *Renewable agriculture and food systems, 22*(2), 86-108.

Barber, D. (2005). Stuck in the Middle. *The New York Times*, November 23, 2005.

Barrett, C.B. (2010). Measuring Food Insecurity. *Science, 327* (5967), 825-828.

Berry, R.A., & Cline, W.R. (1979). *Agrarian Structure and Productivity in Developing Countries*. Baltimore: Johns Hopkins University Press.

Beru, N., & Salsbury, P.A. (2002). FDA's Produce Safety Activities. *Food Safety Magazine, 2*, 13-19.

Boserup, E. (1965). *Conditions of Agricultural Growth*. Chicago: Aldine.

Boyle, M. (2005). A Recipe for Disaster: Scientists race to build tools to defend the food supply from terrorism. But will food companies buy them? *Fortune*, November 14, 2005.

Cameron, G., & J. Pate. (2001). Covert Biological Weapons Attacks Against Agricultural Targets: Assessing the Impact against US Agriculture. *Terrorism and Political Violence, 13*(3), 61-82.

Chapman, D., & Barker, R. (1991). Environmental Protection, Resource Depletion, and the Sustainability of Developing Country Agriculture. *Economic Development and Cultural Change, 39* (4), 723-737.

Clark, C. (1940). *The Conditions of Economic Progress*. London: Macmillan.

Cline, W.R. (2007). *Global Warming and Agriculture: Impact Estimates by Country*. Washington, D.C.: Center for Global Development, Peterson Institute for International Economics.

Desai, G.M., & Gandhi, V. (1990). Phosphorous for Sustainable Agricultural Growth in Asia: An Assessment of Alternative Sources and Management. In: *Phosphorous Requirements for Sustainable Agriculture in Asia and Oceana*. Los Banos, Phillapines: International Rice Research Institute.

Dimitri, C., Effland, A., & Conklin, N. (2005). The 20th Century Transformation of U.S. Agriculture and Farm Policy. *Economic Information Bulletin, 3*.

Eicher, C.K. (1994). Building Productive National and International Agricultural Research Systems. In: Ruttan, V.W. (Ed.). *Agriculture, Environment and Health: Toward Sustainable Development in the 21st Century* (pp. 77-103). Minneapolis: University of Minnesota Press.

Eicher, C.K., & Staatz, J.M. (1998). *International Agricultural Development, 3rd Edition*. Baltimore: The Johns Hopkins University Press.

FAO. (1995). Agriculture, Food and Nutrition in Post-emergency and Rehabilitation. http://www.fao.org/DOCREP/003/V5611E/V5611E00.HTM

FAO. http://www.fao.org/news/story/en/item/47733/icode/

FAO. http://www.fao.org/worldfoodsituation/foodpricesindex/en/

FAS, Foreign Agricultural Service. (2001). *The World Trade Organization and U.S. Agriculture*. FAS online, Fact Sheet, March.

FDA. (2003). *Risk Assessment for Food Terrorism and Other Food Safety Concerns.* Center for Food Safety and Applied Nutrition, U.S. Food and Drug Administration, October 7.

Food & Water Watch. (2008). *The Poisoned Fruit of American Trade Policy: Produce Imports Overwhelm American Farmers and Consumers.* www.foodandwaterwatch.org

Fuglie, K.O., MacDonald, J.M., & Ball, E. (2007). Productivity Growth in U.S. Agriculture. *EB-9*, U.S. Department of Agriculture, Economic Research Service. September.

Hayami, Y., & Ruttan, V.W. (1985). *Agricultural Development: An International Perspective*. Baltimore: Johns Hopkins University Press.

Henderson, D.A. (2001). Biopreparedness and Public Health. *American Journal of Public Health, 91*, 1917-1918.

Jerardo, A. (2003). Import Share of U.S. Food Consumption Stable at 11 Percent. USDA, *FAU-79-01*, July.

Kuznets, S. (1966). *Modern Economic Growth.* New Haven: Yale University Press.

Lever, L. (2013). Average Romanian Household Income Was EUR 540 Per Month in Q3 2012. *Romanian-Insider.com*, January 9.

Lobell, D., & Burke, M. (2010). *Climate Change and Food Security: Adapting Agriculture to a Warmer World.* Dordrecht: Springer.

Lowder, S. K., Skoet, J., & Singh, S. (2014). What do we really know about the number and distribution of farms and family farms in the world? Background paper for The State of Food and Agriculture 2014. *ESA Working Paper No. 14-02*, Rome, FAO.

Maass Wolfenson, K. D. (2013). Coping with the food and agriculture challenge: smallholders' agenda. Preparations and outcomes of the 2012 United Nations Conference on Sustainable Development (Rio+20). FAO. July. http://www.fao.org/fileadmin/templates/nr/sustainability _pathways /docs/Coping_with_food_and_agriculture_challenge__Smallholder_s_age nda_Final.pdf (Accessed on Sept. 2, 2014).

Mendelsohn, R., & Dinar, A. 2009. *Climate Change and Agriculture: An Economic Analysis of Global Impacts, Adaptation and Distributional Effects*. Cheltenham, U.K.: Edward Elgar.

Polimeni, J. M., Iorgulescu, R. I., & Bălan, M. (2013). Food Safety, Food Security and Environmental Risks. *Internal Auditing & Risk Management, 1*(29), 53-68

Pritchett, L., Spivack, M. (2013). Estimating Income/Expenditure Differences across Populations: New Fun with Old Engel's Law. *CGD Working Paper 339*, Washington, DC: Center for Global Development.

Rosen, S. & Shapouri, S. (2009). Global Economic Crisis Threatens Food Security in Lower Income Countries. *Amber Waves*, *7*(4), 38-43.

Runge, C.F., Munson, R.D., Lotterman, E., & Creason, J. (1990). *Agricultural Competitiveness, Farm Fertilizer, Chemical Use and Environmental Quality*. St. Paul, Minnesota: Center for International Food and Agricultural Policy, University of Minnesota.

Ruttan, V.W. (1994). Constraints on the Design of Sustainable Systems of Agricultural Production. *Ecological Economics*, *10*, 209-219.

Singh, S.D. (2011). Absorbing the Food Shock of 2011. *Bloomberg Businessweek*. March 3.

Staatz, J.M., & Eicher, C.K. (1998). Agricultural Development Ideas in Historical Perspective. In: Eicher, C.K. and Staatz, J.M. (Eds.). *International Agricultural Development, 3rd Edition* (pp. 30). Baltimore: The Johns Hopkins University Press.

Timmer, C.P. (1998a). The Agricultural Transformation. In: Eicher, C.K. and Staatz, J.M. (Eds.). *International Agricultural Development, 3rd Edition* (pp. 113-135). Baltimore: The Johns Hopkins University Press.

Timmer, C.P. (1998b). The Macroecomics of Food and Agriculture. In: Eicher, C.K. and Staatz, J.M. (Eds.). *International Agricultural Development, 3rd Edition* (pp. 187-211). Baltimore: The Johns Hopkins University Press.

U.S. Census Bureau. (2001). *Census 2000 Brief: Population Change and Distribution*, April 2001.

USDA. *Foreign Agricultural Service*, www.fas.usda.gov/ustrade

van Bath, S.H.S. (196). *The Agrarian History of Western Europe, A.D. 500-1850*. London: Edward Arnold.

Wheelis, M. (1999). Outbreaks of Disease: Current Official Reporting. Bradford (UK): University of Bradford, Department of Peace Studies, *Briefing Paper No. 21*, May 20, 2002.

Wheelis, M., R. Cassagrande, & L.V. Madden. (2002). Biological Attack on Agriculture: Low-Tech, High-Impact Bioterrorism. *BioScience, 52*(7), 569-576.

Wolfe, S. (2014). Why Americans Spend Less of Their Income on Food Than Any Other Country. *GlobalPost*, May 27.

World Bank. (1982). *World Development Report 1982*. New York: Oxford University Press.

World Bank. (1987). *World Bank Experience with Rural Development: 1965-1987*. (Operations Evaluation Study 6883). Washington, DC: World Bank.

WHO. (2002). *Terrorist Threats to Food: Guidance for Establishing and Strengthening Prevention and Response Systems.* Food Safety Department, World Health Organization.

WHO and WTO. (2002). *WTO Agreements & Public Health.* Joint Study by the World Health Organization and World Trade Organization Secretariat, 62-63.

Chapter 3

The Threat of Soil Erosion and Sustainable Agriculture

Abstract

This chapter briefly examines soil erosion as one of the most silent but equally dangerous environmental threats and presents different approaches to sustainable agriculture before providing a case study of community supported agriculture in the chapters that follow. Obviously, we feel that community supported agriculture is an excellent choice for local, sustainable agriculture, as it can be used as a centerpiece for rural economic development. However, even the biggest proponents for this approach would state that it cannot be the only sustainable form of agriculture used to meet the current and future food needs of the world's population. This chapter will also provide a concise outline of some of the future problems that we consider to be the most important in regards to agriculture and what we propose as possible solutions.

3.1. Soil Loss, 'Peak Farmland', and the Need for Soil Conservation

Before presenting different approaches to sustainable agriculture it is important to say a few words about soil degradation, the silent but extremely dangerous threat for the future of humanity the way we know it. As part of the global effort to fight soil loss, three sustainable agriculture practices for soil conservation will be discussed.

3.1.1. The Dangerous Mixture 'Peak Farmland' – Soil Loss

As stated in the previous chapters, the current food problem is not one of production but one of distribution. The human race has done an excellent job of converting land into agricultural production space. This land conversion has led, in time, to the problem of 'peak farmland' (Ausubel et al., 2013). Land is a finite resource and the best land has been already assigned to agricultural production. Reaching 'peak farmland' pushes a shift of agricultural production to marginal land. As a result, ecosystem services are lost resulting in irreversible environmental damage. This change in landscape leads to a loss of resilience as the number of species in an ecosystem is diminished. At the same time that land was becoming uniform in use, agriculture was going through a similar transformation. Production in agriculture went from that of a large variety of crops to monoculture; typically focusing on corn, wheat, soybeans, cotton and rice (Foltz et al., 1993; Cook, 2006; Lehrer, 2008; Campiche and Harris, 2010; Lin, 2011; Chauhan et al., 2012).

Some may wonder why this is a problem. Throughout history mankind has risen to challenges through creativity, knowledge, invention, and technology. Advancements in communication, planting approaches, weather forecasting, and global positioning systems have assisted greatly in extending the productive capacity of the land. Furthermore, the solution to increasing population, and the associated food consumption needs, has been to produce more by using manmade nitrogen chemicals, irrigation, and mechanization.

However, will these solutions continue to meet the needs of a global population that is expected to reach roughly nine billion people in the year 2050 (United Nations, 2004)? Approximately one-third of the world's land not covered by ice is used for cropland or grazing and if logging is included the percentage increases further (Alley, 2011). Despite this large amount of land under cultivation, there is real concern whether enough agricultural production can occur given the ecological damage that has been caused, the human health issues associated and the projected droughts and floods caused by climate change.

3.1.2. Three Practices for Soil Conservation in Sustainable Agriculture

There are many practices used for soil conservation depending on the geography and use of the land under consideration. Given the focus of our book, we choose to focus on three of the most important for sustainable agriculture: conservation tillage, the use of cover crops and crop rotation.

After more than ten thousand years of plowing, the debate around *conservation tillage*, as no-till or minimal-tilling farming, became more intense after World War II. These practices are more and more used as a sustainable type of soil management with environmental benefits (Lal et al., 2007; Montgomery, 2007; Derpsch et al., 2010). Tilling the soil is broadly used in organic farming to control weeds and to aerate the soil, helping to prevent plant disease. However, there are those that question the viability of tilling. Soil erosion could occur due to loose soil, while the mixing of soil types (topsoil mixed with less fertile soil deeper in the ground) and the use of heavy machinery can compact soil or hardpan which affects drainage and plant root development. Opponents also point out that tilling can damage the root structure of existing plants and can bury weed seed which could encourage weed growth. However, no-till or minimal-till farming drastically reduces environmental pollution because tractors or other machinery, as well as manmade synthetic inputs, are used sparingly if at all.

A very different method from no-till or minimal-till farming is farming with *cover crops* during part of the production cycle. Cover crops are an age-old sustainable approach, well over a thousand years old. A cover crop is a plant grown on a plot of land with the intention of plowing the crop under to increase the fertility, organic composition, and pH level of the soil. Effectively, the farmer is rotating fields and "resting" land to eliminate fertility loss. Cover crops, through their root system s, also help water penetrate the soil and reduce erosion. A variety of cover crops can be used depending upon the objective of the farmer. Additionally, a cover crop can be used to reduce pests, augment the flavor of a crop in an adjacent plot, or provide grazing land for animals. The grazing option has the added benefit that animal manure will be mixed into the soil when the field is plowed, acting as a natural fertilizer. Moreover, some types of cover crops are turned under the soil shortly before planting thereby providing "green manure" for the crop planted on the surface. The turned under decaying crop becomes food for the newly planted crop on the surface.

Often used in conjunction with cover crops, *crop rotation* grows different crops on the same plot of land over time. Crop rotation differs greatly from the industrial approach which commonly grows the same crop on the same plot of land year after year, which, as stated earlier, promotes pests, such as corn borers, and depletes the soil of nutrients. In contrast, crop rotation helps to prevent pests because growing different crops in succession on the same plot of land reduces their populations. Since the crops are grown in succession the nutrients in the soil are increased. For example, alternating corn and soybeans plantings reduces the need for fertilizers because soybeans fix nitrogen into the soil (Union of Concerned Scientists, 2008). Often, crop rotation is used with other sustainable agriculture methods.

Many people think of organic agriculture as the ultimate form of sustainable agriculture. However, there are many sustainable agricultural systems and approaches, some of which are introduced in the next section.

3.2. Approaches to Sustainable Agriculture

The following subsections will briefly review some of the most important, for the future, sustainable agricultural approaches: organic, biodynamic, soilless (hydroponics and aeroponics), free-range, rooftop, and vertical farming.

3.2.1. Organic Agriculture

Arguably the most well-known form of sustainable agriculture is organic farming (Padel, 2001; Scialabba and Hattam, 2002). Nevertheless, sustainable agriculture is related to but not necessarily the same thing as organic farming.

Like sustainable agriculture, the definition of organic farming means different things to different people and organizations. The United States Department of Agriculture has a set of guidelines, as outlined in the National Organic Program, as does the European Commission (Organic Farming website). While the criterion of both these regulating bodies is considered by many to be the gold-standard in organic farming, they are not global guidelines.

However, most would agree that organic farms that engage in the following practices are in fact sustainable:

- Avoiding the use of manmade synthetic fertilizers or pesticides in crop production;
- Avoiding the use of growth hormones and antibiotics in farm animals;
- Providing animals raised on the farm with feed grown without the use of manmade synthetic fertilizers or pesticides;

- Encouraging natural methods that balance carbon and nitrogen levels, for example, allowing micro-organisms to break down organic waste-products in compost piles;
- Maintaining a healthy ecosystem by maintaining biological diversity.

The yields of organic agriculture are comparable with the yields of conventional agriculture (Seufert et al., 2012) and there is already a strong research interest in discussing the conversion to organic farming from intensive farming (Lamine and Bellon, 2009; Lamine, 2011; Sutherland, 2011; Läpple and Rensburg, 2011; Mzoughi, 2011; Läpple, 2013).

3.2.2. Biodynamic Agriculture

Biodynamic farming looks at the farm as a complete living organism (Paull, 2006; Paull, 2013). In biodynamic farming crop and animal production occur simultaneously so the manure from the animals can be used to feed the crops and the plants are used to feed the animals; everything on the farm is interdependent and waste material is minimized.

Biodynamic farming is very sophisticated, as the farmer must have an extensive knowledge of the interactions on the farm because problems with plant or animal growth are analyzed and then addressed with the health of the farm in mind rather than immediately applying nutrients, hormones, or antibiotics to increase the yield. Since the farm is considered a complex system, the farmer must be able to balance all of the interactions between the terrain, the animals, nutrient inputs, water-use, climate, and farming techniques to bring out the optimum in production and vitality of the land (Ryan, 1999; Carpenter-Boggs et al., 2000; Fließbach and Mäder, 2000; Mäder et al., 2002). Crop rotation is particularly important in biodynamic farming.

Due to this integrated, knowledge-based approach to farming, given the proper combination of plant varieties, produce is generally healthy and of high quality, with enhanced flavor and nutrients for consumers. Deciding what to grow on the farm is a complex decision process because the farmer must consider the economics and marketing of growing certain crops and animals, weigh environmental issues, human capital, climate, method of growing, and soil quality.

3.2.3. Soilless Agriculture

As hydrological patterns change and droughts are more prevalent, as soil nutrients continue to be diminished, and pests become more resistant to pesticides, alternatives that do not rely upon the land will be necessary. Therefore, food will have to come from new methods that rely less on the land and manmade chemicals and more on sustainable methods. One area of production that will likely have to be developed further in the future is soilless agriculture.

Resh (2012: 31-32) provides an excellent breakdown of the differences, which are summarized below, of both soil and soilless agricultural production. The first major difference between the two production methods is plant nutrition. Plant nutrition for soil-based agriculture varies greatly by location, soil type, and other local conditions. In contrast, soilless agriculture is controlled completely because the quantities of nutrients supplied to the plants are controlled, as are the levels of pH, and the water supply. Furthermore, because nutrients are strictly controlled by the farmer, the number of plants per acre or hectare can be much greater than in soil agriculture which is limited by the nutrients in the soil. Due to these advantages, as illustrated in Table 3.1, soilless agriculture can provide much greater output yields than soil-based agriculture.

Table 3.1. Comparison of Yields per Acre of Soil-based and Soilless Agriculture

Crop	Soil- based Agriculture	Soilless Agriculture
Beans	5 tons	21 tons
Peas	1 ton	9 tons
Wheat	600 lbs	4100 lbs
Rice	1000 lbs	5000 lbs
Oats	1000 lbs	2500 lbs
Potatoes	8 tons	70 tons
Cabbage	13000 lbs	18000 lbs
Lettuce	9000 lbs	21000 lbs
Tomatoes	5-10 tons	60-300 tons
Cucumbers	7000 lbs	28000 lbs

A second difference is the absence of weeds, insects, animals, and soil diseases. Since there is no soil, weeds and soil diseases cannot occur in soilless agriculture. Furthermore, animals and insects, and their effects, will certainly be diminished in soilless agriculture. Obviously with soil-based agriculture insects and animals can potentially harm the crop.

The third major difference, and perhaps the greatest, is in the use of water. Plants in soil-based agriculture require more water because of soil structure, water capacity, and seepage and evaporation. Besides, saline water cannot be used. In soilless agriculture, much less water is required as the amount of water used is strictly controlled and recycling of water and the capture of evaporation is possible. Additionally, saline waters can be used.

The fourth major difference is in the use of fertilizers. Soilless agriculture uses very small amounts of fertilizer and no leaching occurs because of the controlled environment. As detailed on numerous occasions earlier in the book, standard, soil-based agriculture uses large quantities of fertilizers that leach into the ground and are often inefficient due to the way they are applied. As a result, the amount of waste and susceptibility to disease is much greater in soil-based agriculture as compared to soilless agriculture.

The last major difference we will explore is that plants do not need to be transplanted in soilless agriculture. Due to the controlled environment and that plants are in a permanent location, plants in soilless agriculture can reach maturity quickly. Soil-based agriculture, in comparison, typically requires that plants be started in a greenhouse and then are moved once they reach a certain maturity and the outside climate is appropriate. Therefore, plants take a longer period of time to mature then in soilless agriculture.

The leading soilless method is hydroponics although aeroponics will be also presented briefly in the next subsections. Soilless agricultural research dates back to the 1600s, but hydroponics was brought into the forefront by the research of Gericke in the 1930s (Resh, 2012: 26).

Hydroponics

Organic hydroponics is a unique method of growing crops, doing so without soil. While there are some variations of how to grow food hydroponically, the typical growing method is to submerge plant roots into nutrient rich water. The farmer controls the roots of the plant in this approach because the natural environment for a plant, such as soil, is not being used. Hydroponics is a much more difficult way of growing plants because the farmer must have an excellent knowledge of the growing process so that the water, temperature, and other necessities for the plants are controlled for the roots to prosper.

While using this method is more complicated, there are advantages. Since plants are grown in water, minimal land is needed because more crops can be grown in the same area. Moreover, soil degradation is eliminated. Additionally, water use is reduced, as are outside organic inputs and physical labor from the farmer. Therefore, environmental degradation is diminished if not eliminated.

Hydroponics grows food using 70% less water, as compared to standard irrigation grown crops, without the problems associated with soil, such as lack of nutrients, runoff of agricultural pollutants, and erosion (Despommier, 2010: 107). Therefore, another of the advantages that hydroponics has over other methods of sustainable agriculture is that production can occur in very arid climates. Furthermore, the potential amount of food that can be grown is substantial. For example, tomatoes grown hydroponically can yield approximately 150 tons per acre annually (Resh, 2012: 29). Since yields are high, hydroponic crops can be grown commercially. In fact, there are estimates of at least 30,000 acres in Israel, 10,000 acres in Holland, 4,200 acres in England, and 1,000 acres in the United States (Resh, 2012: 28).

Given all the positives of hydroponic agriculture, one might wonder why hydroponics is not more prevalent. The primary reason is the capital costs for developing the proper infrastructure to grow the plants. Necessary equipment for large-scale hydroponic agriculture includes the greenhouse, a heating and cooling system, a ventilation system, the nutrient formula for the plants, and the computer system that controls the environment in the greenhouse and the flow of nutrients to the plants. To hold the plants there must be a container or bucket system established in the greenhouse. Moreover, drain lines that run the water and nutrients over the plant roots must be planned and laid-out.

Lastly, and depending upon how sustainable the farmer chooses to be, a system which captures water evaporation from the plants and recycles it into the water supply for the plants may also be necessary. Such a system would dramatically reduce the relatively small amount of water needed to grow crops.

Aeroponics

Aeroponics is a similar form of agriculture to hydroponics, invented by Richard Stoner while at NASA. However, aeroponics uses approximately 70% less water than hydroponics (Despommier, 2010: 107). In this form of agriculture plants are grown in an enclosed chamber that stimulates rapid plant growth through high levels of humidity without the use of soil. A computer system controls the nutrient laden water mixture that ensures the root system of the plants stay moist for maximum growth. Moreover, because of the containment system, nearly all the water and nutrients can be recycled.

Aeroponics also has a few other advantages. One of the advantages is that seeds and seedlings grow very quickly. Due to the fact that seeds and seedlings are constantly fed water and nutrients in ideal growing conditions they are mature very quickly. Furthermore, many plants can be grown in a very small area reducing the necessity for and impact on land. The result, like hydroponics, is a high output yield with a much smaller environmental impact.

3.2.4. Free-range Agriculture

Up to this point only techniques involving crops have been covered. However, we would be remiss not to include approaches, although far less in number, that involves animals. One operation of raising animals, confined animal feeding operations (CAFOs), confines animals to a small space where they are overfed and injected with antibiotics and hormones to promote fast development and weight gain, which commonly results in too much manure for the land to absorb. Such an approach to raising animals results in antibiotic resistance and numerous possible diseases, such as mastitis, for the animals. In turn, animal disease can quickly turn into human health concerns like *E. Coli.*

In contrast, animal husbandry in sustainable agriculture typically treats animals humanely, allowing them to graze and roam freely. Free-range animals tend to be stronger and healthier because they are able to move freely and graze on grass or plants instead of feed made from corn and animal by-products to which they are not accustomed.

Additionally, free-range animals help to regenerate the land from the nutrients in their manure. Furthermore, the animals are not injected prophylactically with potentially cancerous antibiotics and hormones which then do not get into either the water or food supply when they are excreted by the animals or become a source of contamination when they are slaughtered.

Free-range animals are conducive to small farms because they rely on intensive management and are highly marketable to customers concerned about the social, health, and ethical consequences of commercial animal husbandry (Ikerd, 2008: 168).

3.2.5. Other Sustainable Agricultural Approaches

Other options exist as well. Arguably the best option is *rooftop farming* in urban areas. This idea has been discussed for several years but relatively few of these rooftop farms have come to fruition. Large apartment complexes and commercial buildings typically have flat roofs which are ideal for growing food.

Rooftop farming has several advantages. First, rooftop space is often underutilized and has plenty of sunlight and warmth which would be good for growing food. Second, rainwater can easily be collected and used to irrigate rooftop crops. Third, rooftop farming can assist in the insulation of buildings, reducing heating and cooling costs and, by extension, greenhouse gas emissions. Fourth, the food grown can be used to feed residents of the apartment complex, or if grown on a commercial building can be sold locally or given to those in need. This locally grown food would dramatically decrease the need for transportation, reducing costs and pollution.

However, perhaps the idea that has garnered the most attention is *vertical farming*. Despommier (2010) wrote extensively on this idea. He proposed vertical farming in response to climate change and global population increases. Vertical farming would not be a quick technological but a major transformation of the way agriculture is performed. Vertical farming is possibly the ultimate form of local farming.

Many of the advantages that Despommier lists for vertical farming are the same as hydroponics and aeroponics as these agricultural methods are the most likely to be used in a vertical farming environment. The one advantage that he lists that is slightly different from that of hydroponics and aeroponics is the use of gray water for irrigation. As water becomes scarcer in the future, whether from climate change or population increases, the recycling of water will be important. In fact, cities produce large amounts of gray water from the removal of solids from black water every day (Despommier, 2010: 173).

3.3. THE *GREEN* 'GREEN REVOLUTION'

These agricultural options are highlighted because the status quo in agricultural production cannot continue given the projected population increase, climate change, soil erosion, and water shortages. Different sustainable methods will be required to meet all the future challenges and food demand. Furthermore, as energy supplies decrease while demand increases, a much greater emphasis will have to be placed on locally and sustainably grown food. For example, in the United States alone, farming consumes approximately 20% of the fossil fuels that are used annually (Despommier, 2010: 168). With local, sustainable agriculture, produce will no longer have to be picked prior early and then ripened while transported hundreds, if not thousands of miles. Instead, food will have to be grown where it can be picked when ripe and eaten without ever having to leave the local region. Therefore, there will be no long distances traveled, no refrigeration needed, reduced storage, and much more environmentally friendly practices. The result will be vastly reduced amounts of fossil fuels consumed.

3.3.1. A Possible Solution

To accomplish this goal, there will need to be a new Green Revolution; a *green* 'Green Revolution.' A good solution would be to decouple agricultural production for food from that of agricultural production for energy. Given the political realities that exist in regards to biofuels, one would be unrealistic to think that industrial farms are going to be eliminated completely. Therefore, one solution would be to try to diminish the amount of industrial farms, and therefore the associated environmental degradation, as much as possible. To accomplish this goal, we propose the following.

All food production should be grown sustainably. Agricultural production near large cities should take the form of those described above, using hydroponics, aeroponics, and vertical farms. In these large cities, where space will be at a premium, a vertical farm complex should be located as close to the center of the city as possible. A location near the center of the city will enable food to be easily distributed to all parts of the city. If a center city location is not possible then a series of vertical farms should be developed throughout the city to ensure that there is an adequate supply of fresh food available to the various neighborhoods in the city. Furthermore, additional agricultural complexes can be developed on the edge of cities to provide food to both the urban and suburban regions. These agricultural buildings do not need to be vertical farms since there is more land available on the city edge and horizontal complexes can be built.

For small cities that are not far from rural regions and where farmland is plentiful, community supported agriculture or a similar option should be employed for several purposes. First, CSA will ensure farmers a steady income for farmers. Second, these farmers can use a variety of sustainable techniques to grow the crops so a steady supply of food is available for the populace. Ideally, a mixture of indoor farming and outdoor farming will be used to guarantee the supply of food year round in cold climates and areas where drought and flooding is prevalent. Lastly, this strategy will create economic growth in these rural regions as farm households will have additional disposable income and will demand goods and services. As a result, additional businesses will be built in the rural areas generating economic development and more wealth.

New and improved seed varieties could be used to develop a sustainable agriculture. Edwards and Belay (2003) discuss the very serious issue of land degradation for the case of Ethiopia. They present a project that had addressed the implementation of productive agricultural systems based on ecological principles that effectively manage and use local natural resources. Similarly, in many places around the world, seeds and know-how were developed over thousands of years of farming. As a result, seed varieties have been improved to resist drought and heat and this knowledge could be transferred to assist in sustainable farming in other parts of the world facing climate change effects. The production of an adequate supply of food in regions which are not currently self-sufficient while using little water would bring stability to regions of the world, such as the Middle East (Despommier, 2012: 223).

The strategy we propose would also eliminate the current problem of distribution as the food would be local and quickly distributed. Hydroponics has been shown to be very effective in Swaziland (Scientific American, 2012). In fact, large quantities of water exist underground in Africa (McGrath, 2012) which could be tapped and used sustainably for Africans to grow their own supply of food.

In other parts of the world which are susceptible to monsoons and flooding, indoor farming is an ideal solution because the environmental conditions are controlled. This strategy is a realistic option to obtain a sustainable supply of food to meet the population and environmental demands of the future. However, the unanswered question is how to pay for such an initiative in developing and transitional countries. In these countries, agricultural food corporations will not, with almost certainty, invest in building the infrastructure necessary to produce food indoors. Therefore, a legitimate question is how will developing and transitional countries be able to develop indoor agriculture.

Perhaps the best financing solution is to take the equivalent money in global food aid contributions and designate that aid for building the necessary infrastructure for developing and transitional countries to produce their own food. In 2010, the global community contributed over US $3 billion for food aid (Hanrahan and Canada, 2011). Transferring this amount to infrastructure for indoor farming projects around the world would quickly enable developing and transitional countries to be completely self-sufficient, producing sustainable, fresh food.

3.3.2. The Benefits

As an additional benefit, countries providing food aid assistance could significantly reduce or eliminate their contributions. Given the large debt loads many developed countries savings in the form of reduced or eliminated aid could be very beneficial. Furthermore, the creation of a functioning agricultural system in a developing or transitional country would create economic growth as the infrastructure would need to be built, resulting in jobs and more money in the local economy.

Additionally, there are considerable economic benefits that would result. Beyond the reduction in transportation and production costs, as well as the infrastructure development projects, employment opportunities will exist. People will need to be employed to run the agricultural operations. For example, there will be a need for agricultural foremen to manage the labor growing the food, the farm labor, and workers in complementary industries such as agricultural equipment and local food processing plants. As employment increases there will be a multiplier effect due to increased income in the local economy. Disposable household income will increase as more individuals have good-paying jobs. The disposable income will circulate through local economies as household's purchase more goods and services. As a result, employment in these industries will also increase, leading to additional disposable income leading to economic and employment growth.

Besides the immediate economic, monetary, and environmental benefits of producing local, sustainable food there are also public health benefits. In developed countries, there would be a measurable improvement in public health statistics. For example, the promotion of sustainable, locally grown food would likely result in a reduction in the number of cases of obesity and diabetes because the consumption of low-nutrition, high-calorie processed foods would decrease. Although research (Ames, 1989; Ames et al., 1995; Gold et al., 2001) indicates that there is no connection between pesticide residue on food and human consumption, many believe there is a correlation. Furthermore, there is little research on the consumption of these toxins in small amounts over many years, or of their interactions with other toxins in the body. Therefore, many are convinced of the connection between food and the incidence of cancer.

In developing and transitional countries the public health connection is much more obvious to the public than in developed countries. Arguably the biggest public health issue in developing countries is malnutrition. The production of local, sustainable food would create the opportunity for not only a greater quantity of food to be available but also for a greater selection of food that could provide the necessary variety of nutrients to stave off health issues from a lack of nutrients. Additionally, with the higher quantities of food available for consumption, child mortality rates would likely decrease. Moreover, total deaths and illnesses, such as diarrhea, due to the contamination of the groundwater supply and manmade chemicals in the food produced would be dramatically reduced because sustainable practices do not damage the immediate ecosystem.

Conclusion

This chapter has introduced the silent problem that humanity is facing due to climate change: soil degradation. When combined with population rapid growth, Western-inspired lifestyle change all over the world, 'peak farmland' becomes another threatening 'peak'. The need to feed future population and to adapt to changes induced by the evolving climate calls for local and sustainable agriculture; in other words a new *green* Green Revolution.

References

Alley, R.B. (2011). *Earth: The Operators' Manual.* New York: W.W. Norton& Company, Inc.

Ames, B.N. (1989). Pesticide Residues and Cancer Causation. In: N.N. Ragsdale and R.E. Menzer (Eds.), *Carcinogenicity and Pesticides: Principles, Issues, and Relationships.* Washington, D.C.: American Chemical Society.

Ames, B.N., L.S. Gold, & W.C. Willett. (1995). The Causes and Prevention of Cancer. *Proceedings of the National Academy of Science U.S.A., 92,* 5258-5265.

Ausubel, J. H., Wernick, I. K., & Waggoner, P. E. (2013). Peak Farmland and the Prospect for Land Sparing. *Population and Development Review, 38*, 221–242. doi: 10.1111/j.1728-4457.2013.00561.x

Campiche, J., & Harris, W. (2010). Lessons Learned in the Southern Region After the First Year of Implementation of the New Commodity Programs. *Journal of Agricultural and Applied Economics, 42*(3), 491-499.

Carpenter-Boggs, L., Kennedy, A. C., & Reganold, J. P. (2000). Organic and biodynamic management effects on soil biology. *Soil Science Society of America Journal, 64*(5), 1651-1659.

Chauhan, B. S., Mahajan, G., Sardana, V., Timsina, J. & Jat, M. L. (2012). Productivity and sustainability of the rice-wheat cropping system in the Indo-Gangetic Plains of the Indian subcontinent: problems, opportunities, and strategies. *Advances in Agronomy, 117*, 315–369.

Cook, R. J. (2006). Toward cropping systems that enhance productivity and sustainability. *Proceedings of the National Academy of Sciences, 103*(49), 18389-18394.

Derpsch, R., Friedrich, T., Kassam, A., & Li, H. (2010). Current status of adoption of no-till farming in the world and some of its main benefits. *International Journal of Agricultural and Biological Engineering,* 3(1), 1-25.

Despommier, D. (2010). *The Vertical Farm: Feeding the World in the 21st Century*. New York: St. Martin's Press.

Edwards, S. & Belay, M. (2003). Healing the earth: An Ethiopian story, *LEISA Magazine 19*(4), 12-13.

European Commission. Organic Farming website (Last update: 24/07/2014). http://ec.europa.eu/agriculture /organic/index_en.htm.

Fließbach, A., & Mäder, P. (2000). Microbial biomass and size-density fractions differ between soils of organic and conventional agricultural systems. *Soil Biology and Biochemistry, 32*(6), 757-768.

Foltz, J.C., Lee, J.G., & Martin, M.A. (1993). Farm-level economic and environmental impacts of eastern corn belt cropping systems. *Journal of production agriculture, 6*(2), 290-296.

Gold, L.S., T.H. Slone, B.N. Ames, & N.B. Manley. (2001). Pesticide Residues in Food and Cancer Risk: A Critical Analysis. In: R. Krieger (Ed.), *Handbook of Pesticide Toxicology*. San Diego: Academic Press.

Hanrahan, C.E. & C. Canada. (2011). *International Food Aid: U.S. and Other Donor Contributions*. Congressional Research Service, RS21279, February 11.

Ikerd, J. E. (2008). *Crisis & Opportunity: Sustainability in American Agriculture.* Lincoln and London: University of Nebraska Press.

Lal, R., Reicosky, D. C., & Hanson, J. D. (2007). Evolution of the plow over 10,000 years and the rationale for no-till farming. *Soil and Tillage Research, 93*(1), 1-12.

Lamine, C., & Bellon, S. (2009). Conversion to organic farming: a multidimensional research object at the crossroads of agricultural and social sciences. A review. *Agronomy for Sustainable Development, 29*(1), 97-112.

Lamine, C. (2011). Transition pathways towards a robust ecologization of agriculture and the need for system redesign. Cases from organic farming and IPM. *Journal of Rural Studies, 27*(2), 209-219.

Läpple, D., & Rensburg, T. V. (2011). Adoption of organic farming: Are there differences between early and late adoption? *Ecological Economics, 70*(7), 1406-1414.

Läpple, D. (2013). Comparing attitudes and characteristics of organic, former organic and conventional farmers: Evidence from Ireland. *Renewable Agriculture and Food Systems*, *28*(04), 329-337.

Lehrer, N. (2008). *From competition to national security: Policy change and policy stability in the 2008 farm bill.* ProQuest.

Lin, B. B. (2011). Resilience in agriculture through crop diversification: adaptive management for environmental change. *BioScience, 61*(3), 183-193.

Mäder, P., Fliessbach, A., Dubois, D., Gunst, L., Fried, P., & Niggli, U. (2002). Soil fertility and biodiversity in organic farming. *Science, 296*(5573), 1694-1697.

McGrath, M. (2012). 'Huge' Water Resource Exists Under Africa. *BBC*, April 20. http://www.bbc.co.uk/news/science-environment-17775211 Accessed August 2, 2012.

Montgomery, D. R. (2007). Soil erosion and agricultural sustainability. *Proceedings of the National Academy of Sciences, 104*(33), 13268-13272.

Mzoughi, N. (2011). Farmers adoption of integrated crop protection and organic farming: Do moral and social concerns matter?. *Ecological Economics, 70*(8), 1536-1545.

Padel, S. (2001). Conversion to organic farming: a typical example of the diffusion of an innovation?. *Sociologia ruralis, 41*(1), 40-61.

Paull, J. (2006). The farm as organism: The foundational idea of organic agriculture. *Elementals: Journal of Bio-Dynamics Tasmania, 80*, 14-18.

Paull, J. (2013). Biodynamic agriculture shows steady global growth, Organic News Community, NürnbergMesse GmbH, Germany [Magazine Article]

Resh, H.M. (2012). *Hydroponic Food Production: A Definitive Guidebook for the Advanced Home Gardener and the Commercial Hydroponic Grower, Seventh Edition*. Mahwah, N.J.: Newconcept Press, Inc.

Ryan, M. (1999). Is an Enhanced Soil Biological Community, Relative to Conventional Neighbours, a Consistent Feature of Alternative (Organic and Biodynamic) Agricultural Systems? *Biological Agriculture & Horticulture, 17* (2), 131-144.

Scialabba, N., and Hattam, C. (Eds.). (2002). *Organic agriculture, environment and food security,* Environment and Natural Resources Series No. 4, FAO, Rome.

Scientific American. (2012). Water for Crops. *Scientific American*, August, p. 24.

Seufert, V., Ramankutty, N., & Foley, J. A. (2012). Comparing the yields of organic and conventional agriculture. *Nature, 485*(7397), 229-232.

Sutherland, L. A. (2011). "Effectively organic": Environmental gains on conventional farms through the market? *Land Use Policy, 28*(4). 815-824.

United Nations. Department of International Economic and Social Affairs. (2004). *World Population to 2300.* New York: United Nations Publications.

Union of Concerned Scientists. (2008). Sustainable Agriculture Techniques. http://www.ucsusa.org/food _and_agriculture /science _and_impacts/ science/sustainable-agriculture.html (Accessed March 22, 2010)

USDA. National Organic Program website: http://www.ams.usda.gov/ AMSv1.0/nop (Last Modified Date: 04/30/2014).

Chapter 4

COMMUNITY SUPPORTED AGRICULTURE: WHAT IS IT?

ABSTRACT

One form of sustainable agriculture that can potentially result in economic growth is community supported agriculture (CSA). CSA is a relatively new concept that develops a relationship between a farmer and consumers that has elements of monetary exchange, work commitment, and volunteer opportunities. The structure of CSA is one where consumers share production risk with the farmer. As a result, this relationship enables small, local farmers to remain in business. This chapter explains what a CSA is, the various types of CSAs that exist, and the history of CSA. Then a theoretical concept of supply and demand of CSA is presented to offer an explanation of how the consumer-producer relationship works.

4.1. INTRODUCTION

Community supported agriculture (CSA) is a concept and a paradigm that includes not only a relationship between farmer and consumers, but also contains a coherent set of related ideas, outlined in Chapter 1, that are consistent and novel when contrasted to the mainstream paradigm of industrial agriculture:

- CSA is a relationship and a commitment, between a farmer and consumers (i.e., members).

- Most CSA farms adopt a sustainable approach to farm practices such as organic or biodynamic combined with a long-term relationship to members.
- CSA farms can take on a variety of formats: producing separately or in combination vegetables, fruits, meat, and sometimes milk.
- According to Henderson and Van En (2007: 29), CSA farms are formed in different ways:
  - from an existing farm or group of farms;
  - through institutional initiatives of organizations such as food banks, religious institutions, land trusts;
  - by farmers or groups of consumers.

The past few years have seen a booming interest in this type of agricultural system. As a result, there is a rich literature on CSA which has shown that there are some limitations. The definition offered here is a traditional form used by Roxbury Farm CSA but it should be noted that it is not used by all CSA farms. The approach has evolved over time to accommodate customers and these changes are debated within the CSA community. Many CSAs do no share crop failures but obtain the crops from other farms instead. Also, having members work on farms is less common these days since liability might be a problem and also supervising neophytes might be difficult. Instead, many CSA farms simply have celebration days instead.

4.2. History of Community Supported Agriculture

Most people would agree that the CSA concept is a by-product of cooperative ideals and associations. Cooperative associations on both the producer and consumer levels emerged in the 19th century in response to perceived impersonal markets. Producer cooperatives were short-lived but consumer associations have prospered in the 20th century. Contemporary examples of consumer cooperatives are food co-ops and credit unions. Although CSA is not the same kind of organization as a cooperative, it does share common elements, in particular, personal relationships, membership, a common bond, and participation in some element of the enterprise.

The CSA movement was largely influenced by the biodynamic philosophy espoused by Rudolph Steiner. In 1924, Steiner gave a series of lectures and wrote:

> A farm is healthy only as much as it becomes an organism in itself – an individualized, diverse ecosystem guided by the farmer, standing in living interaction with the larger ecological, social, economic, and spiritual realities of which it is part. (Ikerd, 2008: 217)

Many believe that the CSA concept originated in Japan in 1975 when Yoshinori Kaneko made an agreement, called the *teikei*, loosely translated to "agriculture with a human face", to supply food to ten families (Henderson and Van En, 2007: 258). Many others believe that CSAs originated in Europe, most likely in Germany or Switzerland, sometime in the 1960s and undoubtedly influenced by Steiner's ideals. Although the farmer-consumer relationships were initiated and inspired independent of each other, the CSA movement in the United States seems to be a descendent of the European model with uniquely American innovations. Since then, the CSA model has been modified for different constituencies interested in supporting local agriculture and has spread around the globe.

Perhaps no other country has experienced more CSA growth than the United States. The first CSAs in the United States date back to the Indian Line Farm in Massachusetts in 1985 (Henderson and Van En, 2007: 143) and the Temple-Wilton Community Farm in New Hampshire, also in 1985. There is debate as to which of these farms was first created. However, most of the literature suggests that Trauger Groh's Indian Line Farm was the first CSA in the United States. In either case, the CSA approach quickly spread throughout New England. Then, after some time, CSAs began to spread throughout New York, New Jersey, and Pennsylvania. Eventually, CSAs extended throughout the country, although certain regions are hotbeds for CSAs: Madison, Wisconsin, Minneapolis-St. Paul, San Francisco, Seattle, and the Northeast.

Currently, more than 12,000 farms market products through CSAs (USDA 2012, Table 43). Note that the distribution of CSAs is roughly the same as population densities in the U.S. This is not surprising since the successful CSA farmers need nearby urban populations. The map provided by the website LocalHarvest illustrates that, in general, CSAs tend not to be in direct competition for land with industrial agriculture.

4.3. Features of CSA Farms

Unlike a pure exchange relationship in the market, the farmer-member relationship generally embodies elements of monetary exchange, work commitment, and volunteer opportunities.

Tegtmeier and Duffy (2005) surveyed fifty-five CSA operations located in Iowa, Illinois, Kansas, Michigan, Minnesota, Missouri, North Dakota, Nebraska, and Wisconsin. They found that the typical farmer is 45 years old with 14 years of experience and a college graduate. More than half of the farmers are female, although they had less experience, approximately eight years. The primary reasons for operating a CSA were environmental and social values. The average CSA has 33 members and the farm size is slightly more than 30 acres. The operations are small with most of the labor supplied by family members. However, the net return per acre is $2,467 which is very high.

As with any successful business operation, CSAs can take a variety of forms to meet the needs of the communities interested in supporting local agriculture. CSAs differ by the level of the financial and labor commitments members must make, the decision-making structure of the organization, ownership arrangements, and manner of payment and distribution of the food (Lyson, 2004: 88-89). However, the essential elements of the CSA that link the farmer, the land, and the members in a common endeavor remain the same. Lyson (2004) outlines the four most common forms of CSAs:

- *Consumer-directed CSAs:* A group of consumers (members) create the CSA farm. They then find a farmer who is interested in the CSA model to produce for them. The members, typically in consultation with the farmer, make the decisions about what food will be grown and how, the share price, as well as the degree to which the members will be required to meet minimum labor requirements. Basically, the members hire the farmer who takes direction from them.
- *Farmer-directed CSAs:* The farmer is the prime mover of the formation of the CSA. He/she seeks to affiliate with interest groups intending to belong to a CSA in order to obtain local agricultural products. In this version the farmer creates the CSA and assumes responsibility for the management. If the structure persists over time, members become subscribers and have negligible involvement in the farm operations.

- *Farmer-coordinated CSAs:* Two or more farmers combine their resources and knowledge to produce agricultural products for a group of local consumers. Each farmer specializes in one or more sets of product. To the extent that this approach is directed at satisfying the mutual interests of all concerned, it may be equivalent to a local farm product distribution system. Farmers can meet the needs of consumers better than any single farmer could individually. Unless, members intentionally commit to support these local farmers in ways that extend beyond the market, it more closely resembles a subscription membership with farmers for a bundle of products than a CSA with community support.
- *Farmer-consumer cooperatives:* Farmers and consumers work together to create a CSA from scratch where there is no already existing farm for potential members to join. They combine their resources to purchase land and equipment for the farm to support the CSA. Decision-making on all aspects of the farm are shared jointly by the farmers and consumers including the agricultural method, the ownership and lease conditions of land and capital, the share price, and the work commitment.

Members could buy annual 'shares' in a CSA farm, where 'shares' can be thought of as a membership fee and an opportunity to participate in some capacity beyond the market in the life of the farm. The share price is predetermined before the beginning of the time period, usually a growing season or calendar year, covered by the share agreement. It may be set by the farmer alone or in consultation with representatives of the membership to cover the operating costs of the farm. Therefore, a 'share' is also a formal commitment to share risk with the farmer since the 'share' is purchased well in advance of the growing season so that the farmer can plan the farm season and buy the necessary inputs (i.e., seeds, fuel, packaging materials, etc.) to operate the farm for the year. Since the membership fees are paid in advance, the members are sharing in the risk in the operation of the farm for the season. Members implicitly agree to share in either the bounty or the shortages that might occur during the season. They trust that the farmer will provide maximum work effort to do his/her upmost to deliver on expectations. Another way to characterize the financial arrangement with the CSA farmer is that members provide interest-free short-term financial capital for the farmer to operate for the growing season. This arrangement lessens both the interest paid and marketing costs to the farmer, thereby removing elements of risk and encouraging the farmer to concentrate solely on farming.

Many CSA farms require members to provide some minimum level of work-service at the farm, support in organizing or administrative membership, or help with the distribution of farm products. Some CSAs allow members to make an extra payment in lieu of the work requirement. However, some will argue that buying out of the requirement makes the farmer-member relationship purely as a market exchange.

In exchange for their membership fees and (possible) work commitment, each member receives a share of the output that is grown or produced on the farm and/or affiliated farms. In addition to financial risk that members undertake and the farmer's production risk, the member incurs inconvenience costs such as having to go to a central location at a pre-specified time and date to pick up the share as compared to going to a grocery store to purchase food products at their convenience. Additionally, most CSA members spend more time cooking and preparing food because farm products are generally unprocessed and perishable as compared to commercial food purchased elsewhere. Members may be challenged by receiving a diverse supply of food which they may be unfamiliar with because the variety of produce CSAs grow might include vegetables that are outside of some members' typical food consumption habits. The member is very likely to get produce which they are unfamiliar with as a sustainable farm must grow produce according to the season. On the other hand, to ensure a constant annual revenue stream, the farmer must provide both high quality and quantity of product to insure that members will want to rejoin the farm again the following season. When members alter their lifestyles, more or less, to conform to the production cycle of the farm during the growing season, their experience usually deepens their relationship to the farmer. If this relationship strengthens then the CSA prospers. Those members unwilling to change will not renew their memberships in the future.

4.4. Raison D'Etre for CSA Farms: Members and Market

CSA and agri-food networks are two of the best alternatives for sustainable agriculture because they incorporate associative relationships and social learning (Krabbe, 2013). Current industrial agri-food system is unable to provide equitable access to food, or to offer nutritious, high quality food in a sustainable way. A movement toward community supported agriculture provides an opportunity to meet the global food demand that the neoliberal approach cannot do because it encourages developing countries to grow high export value crops and importing other crops from developed countries. This approach undermines local production and leads to environmental degradation. Therefore, policies that support local, sustainable production, such as modern agricultural science, indigenous knowledge, farmer to farmer networks, and technology for production, distribution and processing are necessary (Nelson and Stroink, 2012).

Similarly, Brehm and Eisenhauer (2008) find that CSA counteracts the global industrial food system. They use data from CSA operations in Central Illinois and New Hampshire to identify the perceived benefits of CSA membership and why people join CSAs, as well as the community effects of a CSA. Their results show that the reasons for joining CSAs are diverse, but environmental concerns and strong community attachment are common influences. One important result from their research is that they found very little geographical differences in the responses from CSA members.

Differently from the previous study, Pole and Gray (2012) used an online survey to determine the reasons why people joined a CSA farm in New York State in 2010. Results of 565 CSA members' surveys showed that people joined a CSA for fresh, local, organic produce. Few respondents indicated community building as a reason to join a CSA or that they get a sense of community from the CSA.

Analogous to the survey that we will discuss in Chapter 7 for Roxbury Farm CSA in New York, Lang (2005) surveyed 204 members of five mid-Atlantic CSA's in 2000. The results of the study show that member satisfaction is positively correlated with alternative agricultural practices, how many times members visit their CSA, how long a member has belonged to the CSA, and the likelihood that a share in the CSA meets the needs of the member. Furthermore, being an older, female, vegetarian that is a working member of the CSA are predictors of satisfaction.

Sharp, Imerman, and Peters (2002) find that rural-urban relations can be improved by fostering social connections or communication between farmers and non-farmers. They performed a qualitative study of a U.S. Midwestern CSA and found that consumers joined a CSA because they wanted to support the local food system and wanted a quality product while producers indicated that there was a financial interest in involvement. They suggest that Extension programs foster increased interaction between farmers and urban residents because this could lead to new markets and social capital being developed, such as cooperation between producers.

On the other side of the Atlantic, Cox et al. (2008) did a case study of a CSA in Scotland to determine why producers and consumers participate. They find that communication is important within direct marketing because information from producers promotes an understanding from consumers that can lead to a long-term commitment and, likewise, consumers can share amongst themselves about their experiences that could alter the producer-consumer paradigm. Brown and Miller (2008) review the literature on farmers markets and CSAs. They find that additional research is necessary to determine the impacts of farmers markets and CSAs, such as the ecological impacts like water and soil conservation practices. Furthermore, they suggest additional research that explores the relationship between selling into a local food system and the impact on the local and regional environment.

It is known that the organic sector has had a long-standing relationship with local markets. The relationship of organic farmers with local markets is examined by Dimitri (2011) who states that selling products locally is relatively small and new, although she found that 65% of organic farmers sell more than half of their products locally. Another approach to marketing of locally grown products is assessed by Hardesty (2008), who focuses the study on colleges, universities, and teaching hospitals. Data from surveys is used to examine the effects of transaction costs and institutional characteristics on locally grown products. The findings show that colleges and teaching hospitals have high transaction costs and a price premium to initiating a locally grown products program. He also found that locally grown products programs were as a result of wanting to do the 'right thing' and for environmental purposes, rather than from a specific request to buy locally.

Labeling products as sustainable is another marketing technique. Onozaka and McFadden (2009) examine its effects and valuation in the marketplace using a conjoint choice experiment as part of a 2008 U.S. survey. Their results show that locally grown has the highest value, which is further enhanced with the addition of fair trade certification. However carbon-intensive local products are discounted more than those from other locations. They also found that negatively valued claims, such as imports and carbon footprint, can be offset by asserting the products are organic and/or fair trade.

Delving more into the marketing realm of sustainable produce, Onozaka, Nurse, and McFadden (2011) examine the behavioral motivations, choice of shopping locales, and values consumers place on the marketing of sustainable fresh produce, such as organic, fair trade, and carbon footprint. Their findings suggest that consumers who identify with improving sustainability value labeling claims higher and that the information enhances consumers' perceived effectiveness potentially increasing the value of these labels. However, they also find that there is no particular way to target the market for sustainable food entirely.

Closing the sustainable CSA production-consumption loop, the link between market forces and the strategic plans of CSA operations is investigated by Meyer (2012). She finds that CSA farms have average performance in terms of customer service and competitiveness, moderate productivity levels, and a weakness in distribution, profitability, and management of member expectations resulting from the farm's goals of perceived superiority of its products and sustainable agricultural practices used. Furthermore, CSAs struggle from competition from farmers markets, retail produce outlets, and other direct-to-consumer farms. As a result, she suggests a strategic plan where CSAs target consumers who share in the mission of the farmer, consider add-on products from other local farms, and to foster a sense of community among members.

4.5. The Economics of CSA Farms

Ravenscroft and his collaborators (2012) explore the theoretical underpinnings of CSAs. They find that the consumer-producer relationship can be understood as gift-based rather than market-based. The balance of supply and demand for CSAs is examined by Flora and Bregendahl (2012) who use a community capitals framework that incorporates natural, cultural, human, social, political, financial, and stocks and flows of assets to examine expectations of producers and consumers in CSAs. They find that producers and consumers that receive more than one of these goods are more likely to continue their associations. In particular, those individuals who identified CSAs as social, political and economic were more likely to maintain and expand their participation.

Galt (2013) develops a political economic framework for CSAs using economic rent, self-exploitation, and social embeddedness with regards to farmer earnings. He uses qualitative and quantitative data from fifty-four CSAs in California to explain the farmer earnings in relation to production characteristics, social embeddedness, and farmer's motivations. Qualitative data from interviews was used in ordinary least squares regression analysis to show that farmer age, number of employees, and type of CSA impact earnings, that farmers are self-exploitive because of a strong sense of obligation to their members, and that income is not a high priority for farmers relative to other values.

In each of the aforementioned types of CSA farms, the farmer will typically develop the budget and growing plan. Members may or may not have input into these decisions, depending upon the type of CSA. The budget is developed to cover all the operating costs based on prior experience, pro-rated capital costs, and compensation structure for the farmer and the other employees, along with a small contingency fund for unforeseen emergencies. The total estimated costs for the farm season are divided by the number of member shares of the CSA to determine the share price for the membership.

Many CSAs are so small that the farmers are dependent upon some type of outside income and/or rent subsidy or a large work commitment from members because the scale of the operation would make the share price otherwise unaffordable. Therefore, the economically successful CSA must keep in line with what members can afford and long term accessibility to farmland.

Even if the farm is growing and fixed costs increase, the CSA has the advantage of having minimal explicit variable costs because the labor and management of the farmers themselves and members' participation can be substituted for some external inputs (Ikerd, 2008: 136). Thus, a CSA farmer using organic methods can internalize some of the costs by substituting farmer and member contributions to farm work and management thereby avoiding some traditional external costs such as hired labor and chemical inputs. While these costs are real social costs for operating the farm, they are not reflected in the monetary costs thereby keeping down the cost of the share price. For example, if weather adversely impacts farm production, the losses can be absorbed by a smaller return for farmer labor, management and members, whereas high-input, high-variable-cost industrial farms are more vulnerable to risk (Ikerd, 2008: 136).

Perhaps the greatest advantage, economically speaking, for the farmer-member cooperation in CSAs is that demand for the product is purely consumer driven. A growing percentage of consumers are willing to pay a premium price for organically grown food and free-range animal products for several reasons. First, many buyers want to know where and how the food was grown or raised. Second, they want to support local products and producers (i.e., local agriculture). And, third, they want to be ecologically responsible by promoting environmentally sensitive products (Ikerd, 2008: 138).

Therefore, the CSA farmer develops a personal relationship with members, seeking input on all aspects of production and distribution as well as educating members about how the agricultural products are produced. In addition, members are encouraged to visit the farm, to interact with the farm staff, and experience an agricultural environment. Communication with members and their visits to the farm are also important because if production levels are down for reasons outside the control of the farmer, like drought or flooding, members are more likely to understand and to accept a smaller share during the crisis. And despite a reduction in the farm output, the members who have an experience of the farm would be more likely to maintain their membership the following year because the shortfall would be seen as temporary. Education about how food is grown is an important aspect of the farmer-consumer dynamic.

Additionally, as more consumers become educated about the globalization and corporate consolidation of conventional food system issues, the more likely they are to support CSAs and the sustainable agriculture movement (Ikerd, 2008: 160). More people have become aware of how food is grown in the industrial agricultural system because of literature like Barbara Kingsolver's book *Animal, Vegetable, and Miracle*, Michael Pollan's *Omnivore's Dilemma*, as well as documentaries about the food industry such as *Food, Inc.* More and more consumers see CSA as minimizing, if not eliminating, problems associated with the industrial agricultural system including environmental pollution, food contamination, farm-worker health, and excessive food processing resulting in a lack of nutritional value.

These farmer and member preferences and behaviors, along with other variables, were used by Polimeni and his collaborators (2006a and 2006b) to develop in-depth theoretical models of supply and demand for CSA. These results provide considerable insight into the producer-consumer dynamic that exists in CSA operations and are presented in the following section.

4.5.1. Discussion of Demand Models

The authors first started by exploring a standard economic approach to developing a microeconomic model of demand for CSA, then extending the model to account for the nuances needed to more closely portray the dynamics of demand. The first approximation in Equation 1 shows that the quantity demanded of CSA shares during a given time period depends on traditional factors influencing the quantity demanded such as share price, the price of substitutes, income of buyers, tastes, and expectations, among other factors unique to the market for food.

$$Q_{vt} = f(P, P_s, P_c, I, N, T, W, E, H) = f(X_1) \tag{1}$$

Where:

Q_{vt} = quantity demanded of CSA shares
P = price of a food bundle from a CSA farm
P_s = price of a substitute food bundle
P_c = price of complements
I = income
N = number of potential consumers
T = consumer tastes and preferences
M = marketing
W = word of mouth
E = expectations
H = personal health reasons

The demand for a CSA farm membership is the demand for food produced by a particular method and more or less local to the consumer. The price of a CSA food bundle (P) is for the duration of membership. The law of demand suggests that the price and number of shares demanded are inversely related. That is, as the price of the share increases, one would expect that the number of shares demanded would decrease. The share price is easily comparable to the price of a substitute food bundle (P_s) from grocery stores, supermarkets, farmer's markets, etc. if the bundle is broken down into its individual elements for purposes of comparison to other outlets. The CSA member can respond to the price of substitutes between sign-up periods for membership by shifting to competing sources for comparable bundles of food. Economic theory suggests that the shares demanded would be directly related to the price of substitutes.

The price of complements (P_c) typically includes all the associated costs of a CSA membership including travel costs to the pick-up site beyond the amount of normal travel and the opportunity cost of the work requirement. These costs are the extra costs beyond the usual costs of buying food and are relatively easy to calculate. However other nonmonetary costs are more difficult to measure. They include the loss of choice in what the consumer receives in the food bundle, opportunity costs associated with having to store and prepare the food received, as well as the opportunity cost in time of travel to alternative food distributors. The shares demanded are inversely related to the price of complements since the extra costs of concomitant items mitigate the benefits of the nominal share price.

The other variables in Equation 1, *I, N, T, E,* and *H*, are typical for microeconomic demand models. Income (*I*) signifies the average household income level in the market. Since a CSA membership is a normal good, income is assumed to have a positive relationship with Q_{vt}. The number of potential customers (*N*) is a proxy for the size of the market; *N* is assumed to be positively correlated with Q_{vt}. *T* is a bundle of consumer tastes and preferences which is assumed to measure a household's utility for fresh, locally and sustainably grown food; it is also assumed to be positively correlated with Q_{vt}. *E* represents the expectations of the members; if the expectations of members are met or exceeded, Q_{vt} is expected to increase. *H* is a set of health benefits that members anticipate receiving from the CSA produced food. Again one would assume a positive relationship to quantity demanded. Marketing, *M*, is expected to have a positive relationship to quantity demanded because these variables increase consumers 'knowledge about CSA products and provide potential members with information by emphasizing all the advantages of belonging to a CSA without emphasizing the disadvantages. On the other hand, word of mouth, W, is somewhat indeterminate in the sense that the experience of members could be positive or negative. These last four variables related to expectations, health, marketing and word of mouth are important to include because independent consumer choice is highly unrealistic, as family members, peers, and neighbors care about each other's consumption choices (Paavola, 2001) especially with respect to food and health.

However, Equation 1 is not a complete demand function because there are several intricacies of a CSA membership that the equation does not capture. For example, most CSA operating in northern regions are not able to produce vegetables for about six months of the year. Hence, members must take their produce demand somewhere other than a CSA farm for the six months of winter. If the members do not learn or value the importance of their membership during the six months of CSA membership, then the gap between their perceived membership benefit and the actual cost of satisfying the six month period food demand will become smaller. Therefore, the probability that the member will rejoin the CSA the following year decreases. In other words, increases in commitment to the CSA for health, environmental awareness, and support of local agriculture in addition to the market of the bundle of food products obtained from the CSA all contribute to the overall perceived benefit of long-term membership.

Lass and Sanneh (1996) found that market prices are not a reflection of the actual costs or benefits of CSA memberships when external costs and benefits are considered. Building upon this result, Equation 2 offers a model that addresses the complexity and richness of the CSA experience:

$$Q_1 = f(X_1, B_1, G_1, S_1, R_1) \tag{2}$$

Where:

Q_1 = quantity demanded for CSA memberships in the first year

X_1 = a vector of the exogenous variables included in Equation 1

B_1 = the labor requirement by the CSA (some CSAs allow this to be bought out which effectively increases the price of membership)

G_1 = environmental awareness

S_1 = social conscience (i.e. want to support local agriculture)

R_1 = risk shared with farmer

Q_1 represents the demand for CSA memberships. X_1 is a set of exogenous variables that were included in Equation 1. The expected relationship with demand was explained previously, and, as such, will not be explored again. B_1 is the labor requirement that the member must provide the CSA. In some CSAs, members can buy out of this requirement, but by doing so they raise the price of their own membership by the dollar amount of the buyout. For these members, one aspect of the CSA experience is eliminated and the CSA share price moves toward a market price as the member is more like a consumer than a member.

The CSA runs the risk of losing its identity if and only if all members were to buy out of their labor requirement. In that case the farmer/member relationship becomes an impersonal subscription farm rather than a participatory CSA farm. The assumption in our model is that the majority of members are required to contribute work as a condition of membership with a few exceptions. Demand is assumed to decrease as B_1 increases.

G_1 represents the utility members receive from environmental awareness. Among others, these factors include support of organic food because it reduces the amount of chemical fertilizers and pesticides released into the food and ecosystem. Environmental awareness also includes the fact that CSA farms contribute to the reduction in carbon emissions because the food-miles, the distance food has to travel from farm to consumers, are minimized. This awareness increases with the campaign of the "locavore" and "buy local" movements. As a result, environmental awareness and the quantity demanded for CSA member shares are positively correlated.

S_1 serves as a proxy for the social conscience of the member. This variable includes a number of factors that contribute to social justice and harmony among people and nature. Additionally S_1 includes the desire to support a local farmer, to promote better treatment for farm-workers, to eliminate laborers' exposure to harmful chemicals as compared to conventional agriculture, to foster more humane treatment of farm animals because their environment more closely approximates their natural environment, and to be part of a local farm where membership means the land will be treated in a sustainable manner. As social awareness increases, the demand for shares will also increase.

Lastly, R_1 captures the risk that members share with the farmer. Farmer risks include both on farm activity and adverse weather events, both of which can be mitigated by planning, but cannot be eliminated entirely. On-farm risks include the breakdown of machinery at critical times that disrupt, planting, irrigation, harvesting or delivery. Each breakdown carries a risk of a loss of production. Moreover, the risk of farm accidents that involve farm-workers or the farmers themselves who have critical skills needed for farm operation might require the hiring of additional paid help, increase the amount of member work contribution, or a reduction in the amount of a share. In addition, extreme weather such as heat, rain, and wind can wipe out a significant part of the planned production. Finally, on an organic farm there is always the possibility that an insect, fungus, virus, or animal could damage a wide variety of crops or cause farm animal sickness or death. While some of the various risks discussed above can by minimized by taking reasonable precautions, none can be entirely eliminated. Hence, as the risk shared with the farmer increases, share demand is expected to decrease.

Since membership levels in a CSA change annually and are dependent on the knowledge and member experience of all the factors discussed above, a dynamic model of demand for CSA shares that incorporates the learning and the changing values of members must be developed. Therefore, the demand for membership for the second year is represented as follows in equation 3:

$$Q_2 = f(X_2, B_2, G_2, S_2, R_2) \tag{3}$$

Where:

Q_2 =quantity demanded for CSA memberships in the second year

X_2 =a vector of exogenous variables included in Equation 1

B_2 = the labor requirement by the CSA (some CSAs allow this to be bought out which effectively increases the price of membership)

G_2=environmental awareness

S_2 =social conscience (i.e. support for local agriculture)

R_2 =risk shared with the farmer

And

$X_2 = f(L_1)$

$B_2 = f(L_1)$

$G_2 = f(L_1)$

$S_2 = f(L_1)$

$R_2 = f(L_1)$

$L_1 = f(V_1, C_1, Z_1)$

Where

L_1 =learning that occurred in year 1

V_1 =time volunteered beyond B_1

C_1 =willingness or openness to change

Z_1 =social learning

The model presented above shows a set of functional relationships contributing to unique farm–specific knowledge during the first year of membership that contributes to adjusted levels of work commitment, environmental awareness, social conscience, and risk shared with the farmer.

The specific factors contributing to learning are the amount of time volunteered to the farm beyond the minimum work requirement, the willingness of the member to change habits of food preparation, eating habits and life style.

Members learn through their experience as a CSA member and learning changes from the first year endogenously impacts all the variables in Equation 2. Some learning may cause members to reevaluate the weight of importance of demand factors they had originally placed on the first year of enrollment. For example, every year of membership after year 1, share price may become less important to members because they learn more about the value of avoiding the hazards of agricultural chemicals and engaging in sustainable farms methods by working at the farm, through farm letters and through the media.

Thus, G_2, S_2, and Z_1 are included in Equation 3. As these variables increase, demand for a second year of membership is expected to increase.

C_1 is assumed to have a positive correlation with demand.

In large part, C_1 and Z_1 are impacted by the interaction the member has with the farmer and other members because, as stated earlier, a CSA membership causes a lifestyle change. The weekly shares that members receive often contain food that is foreign to members and they must learn to clean, store, and prepare the food. The farmer and other members can assist in this through the sharing of information.

Since CSA membership can continue into perpetuity as long as the CSA farm remains in operation, a generalized theory of CSA demand is necessary and presented in Equation 4:

$$Q_t = f(X_t, B_t, G_t, S_t) \tag{4}$$

Where:

Q_t =quantity demanded for CSA memberships in year t

X_t =a vector of exogenous variables in Equation 1

B_t =the labor requirement by the CSA (some CSAs allow this to be bought our which effectively increases the price of membership)

G_t =environmental awareness

S_t =social conscience (support for local agriculture)

And

$X_t = f(L_{t-1})$

$B_t = f(L_{t-1})$

$G_t = f(L_{t-1})$

$S_t = f(L_{t-1})$

$R_t = f(L_{t-1})$

Where: $L_{t-1} = \Sigma L_1, L_2, L_3, \ldots L_{t-1}$

Similar to Equation 2, all the variables in Equation 4 are endogenously impacted by the consumer learning that occurred during the years of membership. Of considerable importance is that each member will have a different demand function based on their years of participation in the CSA farm. Therefore, the total demand for CSA memberships takes the following form:

$$Q_{CSA} = \Sigma Q_1, Q_2, ..., Q_t \tag{5}$$

4.5.2. Discussion of Supply Models

While CSA is largely a demand-driven concept, there must be a supply to meet some, if not all, of this demand. Polimeni and his collaborators (2006b) present a model for CSA supply. The authors first present a traditional model of supply for CSA memberships, as shown in Equation 6:

$$Q_v = f(P, P_k, P_w, P_L, P_M, I, T, J, O, E_S) \tag{6}$$

Where:

Q_v = the quantity of CSA farm memberships, represented as the supply of food bundles

P = the price of a CSA farm membership

P_k = the price of capital

P_w = the wage rate for farm labor and administrators

P_L = the lease rate of farm land

P_M = the cost of marketing and retailing

I = short and long-term interest rates

T = taxes

J = technology

O = opportunity costs to the farmer

E_s = expectations (including expected inflation, P_e)

Conventional supply theory suggests that CSA farmers would provide more CSA memberships at a higher share price. Likewise, a drop in any or all of the prices of capital, labor marketing or labor would lead to more shares offered. These variables included in this model are standard in a supply equation. However, a few of these variables do require some explanation. *T*, taxes, is included in the supply function because taxes for the CSA farmland will vary by ownership and type of business organization. Some farms, for example, are designated as not-for-profit firms which exempts them from corporate taxes, whereas others are limited liability corporations that are subject to corporate tax. *J*, technology, is rarely included as an explicit variable in a supply model but was included in this version because agricultural product grown through a sustainable approach often requires a specific technology to be used on the farm. Conventional theory would suggest that these technologies would only be used by farmers to reduce production costs and to increase output. *O* includes the price of the land used for agricultural purposes which could be sold for a high price for development. Furthermore, farmers that use sustainable practices might be able to increase their profits by producing value-added products or adding some value to the raw agricultural products. E_s is important because the farmer will manage the supply of memberships based upon expected demand and expected price levels. For example, if expected demand is high, then the farmer will supply a large number of memberships or if expected inflation is expected to be high the farmer will adjust membership prices accordingly. All of the variables in this model are assumed to have a negative relationship with the quantity of memberships supplied with the exception of *J*.

Equation 6 is a good beginning because the model takes into account the fact that the quantity of CSA memberships supplied is not determined solely by share price not by the prices of resource inputs. In practice, the CSA farm must first determine the number of shares that will be offered. The number of shares supplied is related to effective demand for memberships and to resources available to the farm. The number of shares supplied is limited by the available physical resources such as capital, technology, and a fixed amount of usable land. Furthermore, the cost of a membership is not determined until the supply of memberships is determined and the share price is set at a level that provides the farmer an income sufficient for a normal standard of living. Therefore, the price of membership is determined in part as a function of quantity supplied, not the reverse, which means that the income for farmer's family rather than profit maximization is generally a major driver in determining the share price in CSA operations. A closer approximation to the true relationship is shown in Equation 7.

$$P_t = f(\pi_t, Q_{St}, P_{Kt}, P_{Wt}, P_{Lt}, I_t, T_t, E_{St,} M) \quad (7)$$

Where:

P_t = the price of a CSA farm membership in year t

π_t = the profit target of the CSA in year t

Q_{St} = the quantity of CSA memberships supplied in year t

P_{Kt} = the price of capital in year t

P_{Wt} = the wage rate for farm labor in year t

P_{Lt} = the lease rate of farm land in year t

I_t = long-term interest rates in year t

T_t = taxes in year t

E_{St} = expectations in year t

M = member labor, required and voluntary, provided to the farm

In Equation 7, *M* is the labor, both required and voluntary, that members provide to the farm. This variable is included because additional member labor reduces farm costs and drives membership prices lower. Required labor, as explained previously, usually entails that a member helps for a set number of hours at either the farm or the share pick-up site. For example, this requirement could be fulfilled by a member setting up the pick-up site; bringing the food into the pick-up site for other members to collect their share. This obligation reduces the labor costs of the farm. Voluntary labor would be members contributing their personal expertise to the farm without the expectation that they be compensated. For example, a member may have mechanical expertise that could be used to maintain or repair farm equipment. This voluntary action helps reduce both labor costs and variable costs, such as maintenance costs in the example provided. The member volunteer is both a consumer and a non-compensated worker where all members, including the volunteers, benefit equally from their contributions to farm production. I_t in Equation 7 measures long-term interest rates, as opposed to *I* in Equation 6 which measures both long and short-term interest rates, because members are interested in both current and future growing seasons.

For sustainable agricultural methods the amount of capital and technology available are not chosen to maximize output irrespective of other goals. This relationship is shown in Equation 8.

$$Q_{St} = f(\overline{H}, K_t, J_t, E_{St}) \tag{8}$$

Where:

Q_{St} = the quantity of CSA memberships supplied in year t

$\overline{H}$ = a fixed number of acres or hectares of farmland

K_t = capital in year t

J_t = technology in year t

E_{St} = expectations in year t

L_t = labor input in year t

Equation 8 shows that the quantity of CSA memberships in year t (Q_{st}) are a function of the size of the farmland ($\overline{H}$), capital (K_t), technology (J_t), member expectations (E_{St}), and labor (L_t) in year t. These equations are ex ante because the price is fixed before the variable costs of production are incurred. After the share price is fixed, any increase in costs will be reflected in a smaller profit because members are not expected to pay more for their shares during the growing season. Rather, the farmer will receive less income than what was targeted if there is an unexpected increase in costs. On the other hand, if, for example, bad weather affects the quantity and/or quality of what is produced, then farmer's income will not be affected in year t but in year t+1 because some members might not renew their membership. Important to note is that negative scenarios are not the only potential deviations from expectations. For example, an unusually bountiful output for the farm enables the farmer to reach the profit target and members to receive shares that exceed expectations.

For a CSA operation π_t is not profit maximization as for large-scale industrial farms. The goal for the CSA operation is to have a healthy farm system (McCabe, 1998). Additionally, the motivation of many CSA farmers and members is to heal the planet while educating the community and implementing a socially just and ecologically sound form of agriculture (Wann, 2001). These objectives may be a necessity because supporting local agriculture is one reason consumers join a CSA (Polimeni et al., 2006a). On CSA farms where members provide labor, the farmer will not have to hire as much labor and can use alternative land-use agreements that increase the income of the farmer (Lass et al., 2003).

Conclusion

As shown in this chapter, CSA takes a very different approach to agriculture; one that has a positive impact on economic development in rural communities as well as on environmental and public health. CSA takes many forms and requires some adaptation of lifestyles for both the members and the farmer. While the history of CSA is not long in the United States, the future looks strong as the demand for this form of agriculture is growing rapidly. Therefore, learning about the history and types of CSA is important. The models of supply and demand for CSA presented in this chapter were developed, in part, from the history of CSA and the forms of sustainable agricultural methods. In the next chapter these models will provide perspective for how CSA operations evolve in one of the largest CSAs in the United States, Roxbury Farm in Kinderhook, New York, which is examined to illustrate the importance of CSA history and sustainable agricultural methods.

References

Brehm, J.M., & Eisenhauer, B.W. (2008). Motivations for Participating in Community Supported Agriculture and Their Relationship With Community Attachment and Social Capital. *Southern Rural Sociology, 23*(1), 94 – 115.

Brown, C. & Miller, S. (2008). The Impacts of Local Markets: A Review of Research on Farmers Markets and Community Supported Agriculture (CSA). *American Journal of Agricultural Economics, 90*(5), 1296 – 1302.

Cox, R., Holloway, L., Venn, L. Dowler, L., Hein, J.R., Kneafsey, M., & Tuomainen, H. (2008). Common Ground? Motivations for Participation in A Community Supported Agriculture Scheme. *Local Environment: The International Journal of Justice and Sustainability, 13*(3), 203 – 218.

Dimitri, C. (2011). Use of Local Markets by Organic Producers. *American Journal of Agricultural Economics, 94*(2), 301 – 306.

Falk, C.L., Pao, P., & Cramer, C.S. (2005). Teaching Diversified Organic Crop Production Using the Community Supported Agriculture Farming System Model. *Journal of Natural Resources and Life Sciences Education, 34*, 8 – 12.

Flora, C.B., & Bregendahl, C. (2012). Collaborative Community-Supported Agriculture: Balancing Community Capitals for Producers and

Consumers. *International Journal of Sociology of Agriculture and Food, 19*(3), 329 – 346.

Galt, R.E. (2013). The Moral Economy Is a Double-edged Sword: Explaining Farmers' Earnings and Self-exploitation in Community Supported Agriculture. *Economic Geography, 89*(4), 341 – 365.

Hardesty, S.D. (2008). The Growing Role of Local Food Markets. *American Journal of Agricultural Economics, 90*(5), 1289 – 1295.

Henderson, E., & Van En, R. (2007). *Sharing the Harvest: A Citizen's Guide to Community Supported Agriculture.* White River Junction, Vermont: Chelsea Green Publishing Company.

Ikerd, J. E. (2008). *Crisis & Opportunity: Sustainability in American Agriculture.* Lincoln and London: University of Nebraska Press.

Krabbe, R. (2013). Community Supported Agriculture and Agri-food Networks: Growing Food, Community and Sustainability? In Q. Farmar-Bowers, V. Higgins, and J. Millar (Eds.), *Food Security in Australia: Challenges and Prospects for the Future* (pp. 129-141). New York: Springer.

Lass, D.A., Stevenson, G.W., Hendrickson, J., & Ruhf, K. (2003). CSA Across the Nation: Findings From the 1999 CSA Survey. *Center for Integrated Agricultural Systems.* October.

Lang, K.B. (2005). Expanding Our Understanding of Community Supported Agriculture (CSA): An Examination of Member Satisfaction. *Journal of Sustainable Agriculture, 26*(2), 61 – 79.

LocalHarvest website http://www.localharvest.org/search-csa.jsp?m&ty=6&co=1&nm= (Accessed on April 6, 2010)

Lyson, T. A. (2004). *Civic Agriculture: Reconnecting Farm, Food, and Community.* Medford, Massachusetts: Tufts University Press.

McCabe, M. (1998). Farming in the 90's: Cultivating Community Roots. *San Francisco Chronicle.* September 11.

Meyer, J. (2012). Community Supported Agriculture: A Strategic Analysis of the Market and A Competency-based Strategic Plan. *Master's Thesis, Department of Agriculture, Food and Resource Economics, Michigan State University*, November.

Nelson, C.H. & Stroink, M.L. (2012). Food Security and Sovereignty in the Social and Solidarity Economy. *Universitas Forum, 3*(2), 1-12.

Onozaka, Y. & McFadden, D.T. (2009). Does Local Labeling Complement or Compete With Other Sustainable Labels? A Conjoint Analysis of Direct and Joint Values For Fresh Produce Claims. *American Journal of Agricultural Economics, 93*(3), 693 – 706.

Onozaka, Y., Nurse, G., & McFadden, D.T. (2011). Defining Sustainable Food Market Segments: Do Motivations and Values Vary By Shopping Locale? *American Journal of Agricultural Economics, 93*(2), 583 – 589.

Paavola, J. (2001). "Towards Sustainable Consumption: Economics and Ethical Concerns for the Environment in Consumer Choices." *Review of Social Economy,* (LIX), 227 – 248.

Pole, A. & Gray, M. (2012). Farming Alone? What's Up with the "C" in Community Supported Agriculture. *Agriculture and Human Values, 30,* 85-100.

Polimeni, J.M., Polimeni, R.I., Shirey, R.L., Trees, C.L., & Trees, W.S. (2006a). The Demand for Community Supported Agriculture. *Journal of Business and Economics Research, 4*(2), 49 – 60.

Polimeni, J.M., Polimeni, R.I., Shirey, R.L., Trees, C.L., & Trees, W.S. (2006b). The Supply of Community Supported Agriculture. *Journal of Business and Economics Research, 4*(3), 17 – 23.

Ravenscroft, N., Moore, N., Welch, E., & Church, A. (2012). Connecting Communities through Food: The Theoretical Foundations of Community Supported Agriculture in the UK. *CRESC Working Paper Series*, The University of Manchester, Working Paper No. 115, October.

Sharp, J., Imerman, E., & Peters, G. (2002). Community Supported Agriculture (CSA): Building a Farmers and Non-farmers. *Journal of Extension, 40*(3).

Tegtmeier, E., & Duffy, M. (2005). Community Supported Agriculture (CSA) in the Midwest United States: A Regional Characterization. *Leopold Center for Sustainable Agriculture, Iowa State University*, January.

USDA. (2014). United States 2012 Census of Agriculture. Summary and State Data. Volume 1. Geographic Area Series, Part 51.

Wann, D. (2001). Organic Farmers Forge Links with Consumers. *Denver Post.* November 25.

Chapter 5

History of Roxbury Farm CSA[1]

Abstract

Roxbury Farm CSA is one of the largest CSA operations in the United States. This chapter presents the history of Roxbury Farm as part of a case study regarding the operations of a large CSA. An explanation of the formation of the farm, the successes of the farm, and the problems of developing the CSA are presented. Then the developments of the farm, including how the farm has grown, and the resulting changes, both positive and negative, are presented. This information is provided to illustrate how a CSA evolves over time and the challenges they face during good and bad periods.

5.1. Pre-Roxbury Farm CSA Formation

Roxbury Farm has undergone several transformations after its initial formation as a CSA. The CSA has used biodynamic production from its beginning based on the background and interests of the farmer. However, the means of production have been gradually transformed from low technology to high. Moreover, the marketing and distribution have shifted from wholesale markets to a subscription member system where delivery sites receive weekly deliveries.

---

[1] *NOTE:* Unless otherwise noted, the background information provided in this chapter came from a series of taped interviews with the Roxbury Farm CSA operator, Jean-Paul Courtens.

The original Roxbury Farm was started in 1990 on farmland near Claverack, New York by Jean-Paul Courtens, hereafter referred to as the farmer, an immigrant from the Netherlands with a background that includes a degree from the Rudolf Steiner School, Warmonderhof, in his native country. The land was made available to him and his wife at the time, by his in-laws. They leased him the land for a small rental fee in return for maintaining the farm buildings and land and doing any necessary improvements. The farmer brought with him three years of experience in biodynamic agriculture at the nearby Hawthorne Valley Farm in Ghent, New York. His experience also included an established relationship with two organic wholesalers that he had been doing business with. His vision extended far beyond simply growing vegetables for the wholesale market. He wanted to establish relationships with the consumers of his produce in order to establish a reputation for the uniqueness of Roxbury Farm as a biodynamic farm.

The farmer's biodynamic approach was, and continues to be, that the land must be treated as a unique living organism, an approach to encourage diversity and balance of all life forms on the farm including, among others, microorganisms, plants, animals, insects, and people. His unique view was based on the qualities of the land and how to nurture it without applying chemical fertilizers or pesticides. In this non-toxic way the land and all life forms on it can become healthier and more resistant to changes in weather patterns, a concern because of climate change. Healthier plants are also more resistant to outbreaks of plant disease and insect outbreaks. Hence, the farmland is not viewed as an equal input among other inputs, but as a living and breathing productive organism with a uniqueness and life of its own interacting with the plants and animals that are dependent upon it. The land, in turn, depends on all the organisms that contribute to building soil fertility.

The farmer nurtures this ongoing process by working to develop the land into rich, high-quality, healthy soil; important because he must grow enough organic produce to financially support his family and operate the farm. In an economic sense, his objective is to maximize the health of the land (biodiversity and sustainability) subject to the constraint of having to make a living for his family.

In that context, the farmer created a plan to optimally utilize all the land at Roxbury Farm based on the topological and soil characteristics. Early in the history of the 30 acres farm he worked on growing vegetables on level land that was best suited to growing produce, while sloped land was reserved for grazing animals. Some basin land remained as wetland while other land was temporarily set aside for cover-crops in order to build up its fertility for future vegetable production. In order to prevent depletion of the soil, each segment of land used for growing vegetables would be taken out of production and cover-cropped for at least one of every three years, a practice used for over a thousand years and often referred to as a three-field system of crop rotation. These methods were used to increase fertility and the long-term sustainability of the farm.

Early in 1990, the farmer started some plants in a greenhouse and worked on renovations on farm buildings and the installation of a rudimentary cooler in an old barn. He obtained a bank loan for equipment needed to get the biodynamic farm operations underway. The farm income the first half of the 1990 growing season was generated by producing organic food for Hawthorne Valley and the Berkshire Morganicus organic wholesale outlets. However, halfway through the season, wholesale prices for the vegetables produced and offered for sale fell to less than the cost of growing the food. This price collapse was brought about by intense competition for market share by large West Coast organic wholesale distributors and could have been catastrophic for the fledgling farm.

New markets had to be found quickly. Fortunately, a relationship formed with a few people in New York City who were looking to develop an association with a farmer to start a small CSA. Additionally, he learned about a newly emerging market for upscale vegetables from a friend and quickly adjusted by plowing under the vegetables he could not sell and grew baby vegetables and other exotic produce to sell at upscale produce markets in the New York City area. Each week, a local farmer would take Roxbury's produce along with his own to the farmer's market in Union Square in New York City. Simultaneously, the Roxbury farmer expanded his operation by selling to several local restaurants and wholesale markets within easy driving distance of the farm. The sales were good but very risky as they were not stable. The farm survived that first year buying time to explore and secure other possibilities for generating long–term, stable revenues for the farm.

The farm reached, by the fall 1990, an agreement with the Omega Institute, an institute for holistic studies in the Taconic Valley of New York, to provide them some vegetables for their food service during the summer when people attended their workshops and short courses. The farmer realized that Omega could provide an element of basic demand and income but a more permanent and stable source of income would be required if he were to survive economically while maintaining his biodynamic ideals. As long as the buyer for the Institute and their chef were satisfied with the quantity, quality and variety of produce from Roxbury, the farmer could depend on them for a substantial order each week in the summer. Fortunately for Roxbury the chef at the Institute welcomed the challenge of developing a menu after he would find out about what produce would be available from the farm for the upcoming week. However, that chef eventually left Omega Institute and the new chef was not as flexible about creating a menu. As a result, the farmer felt that he could not depend on institutional buyers for long-term relationships.

Dependence on relationships with a few institutional buyers, even though he had a personal relationship with the buyers, put the farm at risk because a significant share of the market could be lost when even one institutional buyer changed due to changes in personnel or changing corporate ownership. The ultimate solution was to sell his produce directly to consumers who demanded chemical free food and were concerned about the source of their produce as well as the growing method. Some consumers were willing to pay a premium to establish certainty about the freshness and healthiness of their food, as well as knowing where and how their produce is grown. The farmer also realized that his customer base needed to be broadened to reduce the risk and impact of an individual customer leaving.

However, the farmer had a deep concern of turning agricultural output into a pure commodity. How could the reputation and personal connection to the farm survive the many layers of wholesalers and retailers between the farmer and the ultimate consumer of food from Roxbury Farm? In his view the very nature of farming is dealing with living entities not conformable to the impersonal market. Industrial agriculture, aimed at producing identical products year-round with a long shelf-life, has changed the face of agriculture in order to produce for the market. As a consequence, agricultural practice and technologies evolved to conform to the demands of the marketplace rather than to seek long-term sustainability. His brief, but bad, experience with institutional and wholesale markets left the farmer seeking a mechanism that maintained some personal link between the consumers, the land, and himself.

He had learned from his biodynamic studies that maximizing profits from agricultural production is based on the exploitation and depletion of the land. In the neoclassical model, the land is treated as an inanimate input in the food production process, which amounts to the farmer depleting the intrinsic qualities of the land. This approach fails to take into account the reality that the land is a living organism or the fact that the time horizon for responding to produce markets requires short-term planning and production for immediate sale after harvest. Even with careful planning, produce markets are rife with uncertainty. For example, if there happens to be a bumper crop of the very highest quality vegetable, the market price will be very low. A recurrence of such an outcome could drive even the most productive farmer out of business. In contrast, conserving and building up the land and making a living from agriculture somewhat de-linked from the market, requires long-term planning and sustained long-term economic and social support. This 'long-term' reality motivated the farmer to seek an alternative to the standard 'short-term' market mechanism through which he could avoid this conflict-of-interest that would constantly threaten both the farmer and the land. He knew that in order for small organic farmers to survive, some alternative that embodied a personal relationship to customers would be needed. In short, the farmer and the consumer would associate with each other in some form of long-term mutual support relationship founded on their common set of economic, ecological and social interests.

When the farmer was a student in the Netherlands, he studied the Dutch model of Producer-Consumer Associations. In these associations, produce growers and consumers met once a year to negotiate both the prices and quantities of a myriad of vegetables. Once agreed upon, the producers then developed a production plan for the following growing season to meet the contract conditions. The producers would be assured a sufficient stream of income for the season while engaging in sound agricultural practices benefitting both the farmers and the consumers. Hence, farming enterprises that would otherwise depend on fluctuating prices in wholesale markets would be able to farm without the risk of volatile market prices while consumers would have the assurance of the availability of their preferred food items at privately negotiated prices they could afford.

However, risks remained. For example, weather could impact production resulting in insufficient production which would negatively impact farm income. Thus, producers for the association were engaged in direct to consumer marketing. For the association farmers, production was a personal service to the consumer partly because a contract has been negotiated in advance of the actual farming and partly because the consumers know where their food is coming from and how it is grown.

The association model emphasizes a contractual relationship between consumers and producers giving the farmers reasonable certainty that their farming effort will be sufficient to sustain their families although, this model does not create a *personal* relationship between the farmer and the consumer. With that association model in mind, the Roxbury farmer realized that he wanted to develop some form of contractual relationship between himself and his customers. The seeds of the future Roxbury Farm CSA were planted by a few consumers in New York City in the fall of the first year of farm operation.

5.2. Roxbury Farm CSA: The Beginnings

In 1990 there were far fewer CSAs in North America than there are now, perhaps a dozen at most. Those CSAs were relatively small, labor intensive, low-technology organically farmed operations. In this context, a socially active New York City resident who had heard about the CSA model, envisioned the creation of a rural-urban connection for supporting organic agriculture through the formation of a CSA. He provided the leadership in founding the New York City group's association to what would become the Roxbury CSA Farm. He began with a core group of like-minded individuals who expressed a longing for a reversal of the slow deterioration of the quality of their food and their almost complete alienation from the source and grower of their food. This core group started to investigate opportunities to create a true urban-rural relationship with a specific farm and farmer. They leaned toward the CSA farm model and to biodynamics in particular because these approaches coincided with their hope of supporting a farmer who cared about building up the land's fertility and sustaining the land at a very high level through sound agricultural practices. The benefits would be social, personal, economic (short and long-term), and a healthy, sustainable environment.

The social benefit would be the relationship between the farmer and members who would be committed to facilitate and support each other in the production and distribution of produce. The material economic benefit to prospective members/consumers would be a weekly supply of in-season produce at a prepaid price that they deem affordable, while the farmer would be assured the operating capital for the growing season, similar to the Producer-Consumer Association that was already within the experience of the Roxbury farmer. The new element that was formalized was a working relationship with a particular farmer.

The environmental benefit would be the promotion of diversity on the land that would be dedicated to the production of a wide variety of consumable organic vegetables with farming methods that minimize pollutants from external inputs. The formal mechanism for supporting this kind of relationship with the farmer was a share agreement which established a membership fee through which the members would share in the financial risk with the farmer. The risk included crop reductions due to unanticipated weather or blight at the farm. Members also agreed to provide a few hours of work; whether at the delivery site or at the farm during the growing season.

The core group of potential New York City members had a series of meetings with the Roxbury farmer in the winter of 1990-91 (Hilton, 1998) at which the Roxbury farmer enthusiastically embraced the proposal to form a CSA. He did so for several reasons. First, there is the financial aspect; securing a guaranteed income so he could continue farming. Second, he was looking for some kind of personal relationship with the consumers of his farm produce. Third, his experience with marketing his organic produce led him to a dissatisfaction with the impact of markets on his life. In his words, he "realized that ultimately, in order to be successful, I really needed a personal relationship with the customers and also with the clientele where people really understand the unique things we are doing on my future farm" (Courtens, 2009).

The New York City group had a nearly identical set of goals in terms of ecological concerns, as well as a desire to develop a relationship with a farmer, preferably a biodynamic farmer, and to receive a dependable supply of fresh organic produce. In the spring of 1991, thirty people in New York City joined with the Roxbury farmer to form a small CSA. Weekly distributions of produce to members began in June of 1991 at a New York City farmers' market (Hilton, 1998). The satisfaction and enthusiasm of the New York City members' served as a magnet to attract over fifty more members by the end of 1991 delivery season. The Roxbury CSA was launched.

The farmer wanted to serve the needs of the members in the best and most efficient way while satisfying their mutual objectives. In contrast to other small, labor-intensive CSAs, the Roxbury farmer embraced simple and affordable mechanical technologies that improve the efficiency of the farm operation without compromising the ideals of biodynamic farming. The farmer established a solid, professional relationship with the initial members by matching, if not exceeding, their expectations by producing 400 pounds of high quality in-season produce per share. Additionally, the farmer created a personal relationship with these members by making the deliveries to New York City himself (allowing him to personally meet and talk with members), publishing a newsletter for members, and inviting members to the farm for farm festivals and workdays. As a result, the New York City membership became the first solid core of members of the CSA soon followed by members in Columbia County and the Capital District of New York State. These three core constituencies would become the backbone of the CSA for more than twenty years.

5.3. Roxbury Farm CSA's Expansion

After establishing Roxbury as a CSA farm with New York City membership, the farmer looked for opportunities for expansion to further stabilize his way of farming, as well as to secure a higher income for himself and his family. However, a truly sustainable farm would need to sell locally. He would soon realize that there were opportunities to expand his vision and the CSA operation in his local region of New York State, specifically the Capital District and Columbia County.

5.3.1. The Capital District of New York

During the first year of the CSA in 1991, the Head of the Peace and Justice Commission of the Albany Catholic Diocese contacted the Roxbury farmer because she had begun exploring ways in which the Commission might support local family farms. The first exploratory step was taken by several members of the Commission visiting Roxbury Farm (Scheib, 1994). The Commission members decided to establish a pilot project with Roxbury to recruit Capital District members to the Roxbury Farm CSA. Participation in the pilot program would be accomplished through a series of informational meetings arranged by the Commission at local parishes, both in Albany and Schenectady. At those meetings, the farmer's passion for his work and his vision put a face on the appeal of the Commission to preserve local farms and support sustainable agricultural practices. The organizing effort led to one hundred families from the Capital District becoming members in the CSA by the Spring of 1992. The Commission arranged for two delivery sites in Albany and two in Schenectady close to the members' parishes.

This new Capital District membership more than doubled the size of the overall CSA in its second year. At this point, Roxbury Farm CSA became one of the largest CSAs in the movement. As the membership in the Capital District grew by word-of-mouth, its membership became increasingly diversified away from its almost entirely Catholic roots. By 1993, the Capital District pilot program evolved into another fast growing and durable component of the Roxbury Farm CSA.

5.3.2. Columbia County

A third long term component of the CSA would also emerge in 1992. At the advice of a local doctor's concern for her patients to have a more wholesome diet, a health conscious membership formed in Columbia County, New York. Their pickup site was at the farm, convenient for the local members and efficient for the farm. As a result, the Columbia County members paid the lowest membership fee (no transportation costs). The Columbia County group started with twelve households and grew steadily over the years, helping to create a strong foundation for Roxbury in its early years.

5.3.3. Further Expansion

The mid-1990s was a period of growth for Roxbury CSA. Tireless efforts of the core members in the three regions yielded 650 members by the end of 1995; 230 in New York City, 80 in Columbia County, and 340 in the Capital District (Roxbury Farm Letter, 2001). By then, Roxbury Farm CSA had already developed into one of the largest CSAs in the United States since most CSAs were, then and now, operations serving fewer than a hundred members. While the three core CSA affiliations were initiated for three different reasons, philosophy, religion, and health, they were all motivated by a spirit of cooperation in supporting local agriculture. The New York City group was motivated to advance the philosophy of Rudolf Steiner through their support of a biodynamic farmer. The impulse of the Capital District group was to enhance the integrity of the land while attempting to protect local farmers. And, the motivation of members in Columbia County was driven by a desire for healthy produce and healthy members. To some degree, aspects of these motivations are shared by all the three core groups. These three initial membership bases continue to make up most of the current membership.

5.3.4. The Secrets of Success

The early success of the farm was driven by several factors. First, new members were recruited by word-of-mouth as well as through farmer presentations sponsored by core members. Second, the farm established an education program for apprentices, Collaborative Regional Alliance for Farmer Training (CRAFT), which has become a model for apprenticeship programs nationwide. CRAFT provided a program of apprentice visits to various organic farms where they could learn from the experience of other farmers and working on the farm. The farm benefitted from the apprentices learning, and the farm operation became more efficient and productive. The apprentices would sometimes come back to the farm in successive years further enhancing the efficiency of the farm. Due to increasing efficiency over the years, the cost per pound of produce was systematically lowered each year. Third, the farm actively sought out and implemented appropriate technologies and machinery, such as irrigation guns, planters, and cultivators. The criteria for implementing technology on the farm was that it must reduce the strain on farm personnel, even when the technology did not reduce cost per unit or improve efficiency. Fourth, newsletters, where the farmer educates members about the dynamics of the growing season on the farm, were distributed to members. These newsletters helped solidify the support for the biodynamic ideal by deepening the relationship between the people working on the farm and the members and providing a wider picture of CSAs as a market alternative. The newsletters also provided reinforcement for members who looked for evidence of the positive environmental impacts of sustainable agricultural practices by the farmer, giving members a reason to be proud of financially supporting the farm. Finally, Roxbury Farm became a model for how a successful CSA can be established and sustained. The success and reputation of the farm helped breed other CSAs. However, after eight years of success, new issues and concerns arose from this very rapid growth.

5.4. The New Roxbury Farm

In 1999, personal domestic circumstances caused the farmer to be separated from the farmland in Claverack and have to look for another farm. While he lost his share of the real estate partnership in the farm, he retained the farm equipment, the Roxbury Farm name, and the business. In late 1999, the farmer completed the closing on another farm in Kinderhook, New York, about fifteen miles from the Claverack location. The outstanding issue would be “Will the members be willing to re-enroll in the same CSA but at a different location?” The answer was mixed but not fatal to the CSA as this was less important to the New York City and Capital District memberships, but very much a concern to the Columbia County group. Those members had formed a closer relationship to the Claverack farm because this location was their pickup site. Retaining their membership in the CSA at the new location would require most of those members to travel a longer distance to the new farm, for some an extra 30 miles. However, since the Columbia County group always had fewer members than the other two major areas, the membership loss was a small percentage of the overall membership.

The new farmland was held by the Open Space Institute (OSI), an organization that had been buying land in the Kinderhook Creek basin in order to protect the land from commercial or residential development. As a result, OSI sold the land to Roxbury for less than the purchase price. In exchange, the farmer had to agree to use the land solely for open space or agricultural use. In this way, OSI systematically protected the Kinderhook Creek watershed from development.

The financing of the new farm and related property transactions tested the relationship of the farmer to the CSA members. The farmer, along with the core members of the farm, mostly representing the Capital District and New York City, and a representative of Equity Trust (the trust fund that would finance the purchase of the farm) met with representatives of OSI in the fall of 1999. Equity Trust, with the promise of cooperation and support from the CSA leadership, purchased 140 acres of farmland stripped of development rights for use by Roxbury Farm. An additional 13 acres containing some farm buildings and a farmhouse were eventually sold to the farmer on a limited equity basis. Both transactions ensure that the farmland and the farmer's house will be reserved for farmers in the future. In addition, Roxbury Farm was able to lease another 126 acres of nearby land from OSI and rent a few acres from a nearby farmer for growing organic vegetables during the first year of the move since the new land needed to remain fallow for another year in order to qualify as suitable for growing organic crops. The leased land is contiguous to the Martin Van Buren National Historical Site which is administered by the National Park Service which wanted the leased land to continue in agriculture because Martin Van Buren had farmed the land 150 years before. Fortunately for Roxbury Farm, the land contained an old farmhouse that became apprentice housing. The land also had a set of run-down buildings that were converted into an office, a barn for washing produce, a machine shop, and a greenhouse. These facilities, though old, served the farm well over the next ten years. However, the long-term plan of the National Park Service includes repossession and demolition of the buildings sometime after 2020.

Early in the year 2000, tractors, equipment, and the business part of the CSA were moved onto the new property. After a great deal of hard work rehabilitating the buildings and putting the greenhouse into production, the farm quickly resumed operation. Most of the produce grown that first year was planted on the leased and rented land because of the presence of residual agricultural chemicals on the newly acquired Equity Trust land. To be safe and consistent with organic farming standards and to ensure that the new farm would be chemical free, the farmer left this new farmland fallow in order for all the chemical residuals to dissipate and breakdown.

The farmer describes the challenge of starting over again at the new farm:

> "When we bought the farm it was all planted in corn or potatoes. We found that almost every inch of the land had been tilled and planted to maximize its production. There were ruts from tractor tires two to three feet deep in sections of fields that are too wet to grow crops in. The farm was treated like an object and the crops were seen as commodities."

During the early transition, the core committee of members, in conjunction with Equity Trust and the farmer, launched a Capital Campaign to begin raising money on behalf of Equity Trust to cover the cost of the land. Additionally, Roxbury hired an experienced biodynamic farmer to run the farm operation so that the Roxbury farmer could be directly involved in the fundraising. This resulted in the farm operating at a loss that year and the next. The transition to the new farm actually elicited members to increase their level of voluntary contribution, supporting the farm both in terms of volunteering time and talent, as well as contributing money to the campaign. As evidence of the voluntary efforts of members during the transition, the Roxbury Farm Letter (Week 14, 2000) reported that the combined volunteer labor contribution to the farm was equivalent of two to three full-time jobs. Despite that positive development, the farm struggled in the transition years 2000 and 2001 as 85 of 650 members, mostly from Columbia County, did not renew their memberships. At the same time, the farmers continued to pay attention to the business side of farm operations and earned recognition from the Columbia County by achieving the Chamber of Commerce *Commitment to Agriculture* Award for two consecutive years.

In 2001, a second farmer, Jody Bolluyt, joined the farm as a co-farmer. She brought with her practical farm experience, management skills, as well as a formal education in botany and biology. Now the farm had two full-time professional biodynamic farmers. Thanks to its very good reputation in the arena of biodynamic farming, during the transition, Roxbury Farm was able to recruit veteran apprentices both experienced and capable of taking responsibility for certain specialized functions, such as greenhouse operations or equipment repair, thus adding strength to the farm expertise.

By the end of 2001, members had raised over $200,000 for the Capital Campaign in support of Equity Trust's good-faith purchase of the farm. These contributions led to a formal arrangement in the form of a lease that would secure the future of the farm at the new location, as well as lead to new arrangements that would protect farmland and make that land affordable to future farmers. The formalization of the long-term lease agreement between Roxbury Farm and Equity Trust took five more years to complete because of the complexity of developing a new model of farmland acquisition and ownership.

In 2004, OSI sold an additional 100 acres of farmland to Roxbury, again without development rights. The land was affordable for Roxbury because OSI once more absorbed the cost of the development rights which otherwise would have been prohibitively expensive for the farm, but would have been attractive for a developer.

As part of the expansion process, by the end of 2004 the farm had added delivery sites and members from Westchester County. Consequently, membership in the CSA farm increased to 800.

5.5. Roxbury Farm 2005 to Present

The expansion to include Westchester County added stresses and opportunities to every aspect of the farm operation, including the scale of the greenhouse, tilled acreage, washing and packing facilities, truck capacity for delivery and record keeping on the farm. The response was to add more experienced apprentices and farm workers, and to increase farm technology, including a larger irrigation reel, an additional cultivation tractor, a plastic mulch lifter, and large storage bins for fall crops. A second expansion would see the introduction and the gradual increase in the number and variety of farm animals. The primary purpose of introducing an animal presence was to move closer to the biodynamic ideal of creating balance in the farm's ecology. The secondary purpose was to generate additional income from the sale of meat to members. The farm added lamb and beef in 2004 and then pigs a year later to make use of pasture areas in a productive way. Although expensive due to the high cost of organic feed to supplement pasture feeding and the cost of meat processing, the meat shares typically sold out.

In 2006, due to a very bountiful harvest, the transmission on the delivery truck stop working because the truck was continually loaded at or beyond weight capacity. The farmer explained to the members that a share price increase was necessary to pay for a new refrigerated truck for the 2007 season. The farm also provided an option for members to make multi-year share agreements in order to raise enough money to purchase a new truck without resorting to credit. In addition, the farmer supplemented the revenue from the membership fee increase by selling bulk storage vegetables to those members that wanted them. After repairing the delivery truck in the winter season, a decision was made to limit the number of shares to 1,000 in order to best serve the members. Renewal notices and enrollment forms for membership are sent out in November or early December and members have preference to renew their membership as long as they send in the form and their membership fee by January 15. After that date, open slots at various pickup sites are made available to those who would like to join the farm. Since 2005 there has been a waiting list for membership at each of the delivery sites.

The 2007 season saw an expansion of the number of animals integrated into the farm in a move toward more biodiversity and to expand the meat share offered to members. In contrast to industrial animal feeding operations, the farmers describe the way animals are treated at Roxbury:

> "Any of our animals should have a life that resembles the environment of its wild counterpart (safe from predators); we don't treat it like a pet; and we see it as food, good food, honest food, raised on a wholesome diet, preferably grass." (Roxbury Farm Letter, 2007)

By 2007, Roxbury became known as a pioneer in the Northeast and in the United States in its farm practice. Visitors from across the world have visited Roxbury Farm, including media from Japan, South Korea, and the Netherlands to learn about the Roxbury Farm model. Several articles about Roxbury Farm have also appeared in local and national publications, including the New York Times.

Despite the fact that membership had grown to over a thousand people, the oil price spike in the summer of 2008 hit the farm hard. The spike created an unexpected financial burden to the farm. The farmers communicated this situation to the members in the farm letter indicating that:

> "...we are definitely hurting from the rising gasoline prices. Our fuel expenses so far are at 200% of last year's rate. We also had to give our workers a raise, as they simply can no longer live on the wages we paid in 2007 with normal inflation. Inflation for folks like us is more like 15%, as most of our income goes to heating oil, gasoline, food, and insurance. We

might need to consider the introduction of a voluntary surcharge if we get in trouble financially as the season goes on." (Roxbury Farm Letter, 2008)

By the end of the season, many concerned members supplemented the farmer's income by sending in a member share adjustment of $25. The oil price spike was shortly followed by a hail storm that would temporarily reduce the weekly share as many crops were damaged. Here again, members expressed their support and willingness to share the risk with the farmer. Good weather later in the summer offset the shortages in the weeks following the storm. Also, a fruit farmer that provided the fruit share to the farm experienced so much storm damage that he lost his entire wholesale market. However, his blemished fruit was made available for distribution to members, reducing his financial loss without any loss of nutrition to the members.

By 2009 the farm focused on improvements in farm practice, food safety issues, and animal husbandry. The farm practice shifted toward using green manure and hay mulch as part of its sustainable farming practice:

> "The farm is about 300 acres, about 90 of which are devoted to vegetable crop production, 100 to hay, and about 50 to permanent pasture. To maintain fertility on the vegetable land, we import compost and plow under a significant amount of green manure. Green manure is basically a crop that is specifically grown to be plowed under in order to provide fertility to what we call a cash crop (a crop that is exported off the farm)". (Roxbury Farm Letter, 2009)

This practice complements the expansion of sheep, pigs and cattle on the farm because it separates the vegetable production and the animal operation. The washing barn and packing operations saw improved floors, lighting and washable surfaces. Lastly, the farm received the Animal Welfare Approved label for the animal operation. By 2011, the farm had expanded to 16 delivery sites and 1100 member shares. The Capital District continued to be the largest component of membership with delivery sites in Albany, Schenectady, Delmar, and Troy, and Menands.

Conclusion

This chapter presented the history of Roxbury Farm CSA. How the farm came to fruition, its struggles and successes, as well as some elements of the financial aspects of the farm were presented to show the evolution of a successful CSA operation. As was shown, Roxbury Farm CSA began meagerly but over time became a very successful enterprise.

However, the real success of Roxbury Farm CSA has been in the membership. The original members took a large risk on a largely unproven farm operation. Their belief in the farmer and the biodynamic philosophy helped the farm through the early years. The early members provided revenue for the farm, income for the farmer, free marketing by word-of-mouth, and labor. As a result, the farm grew.

The growth of the farm brought both successes and challenges. Successes included expansion in the size of the farm, equipment, workforce and apprenticeship, and revenue. However, the expansion also brought challenges in how to handle the growth and keep the membership happy. Furthermore, the expansion, while a positive in overall revenue for the farm, increased costs that the farm was unable to absorb at times. Fortunately, the membership supported the farm in times of financial hardship.

This chapter illustrates all the stages of growth of a successful CSA and provides important information for people to understand how to develop a CSA. Arguably the most important aspect of a CSA operation is the membership. As a result, CSA operations should spend considerable time cultivating relationships with members, such as what the Roxbury Farm CSA farmer did with his membership. The next chapter builds on this information by providing details about the financial history of the farm.

REFERENCES

Courtens, J.P. 2009. Interviews during fall 2009.

Hilton, J. 1998. In the Beginning: An Exercise in Practical Idealism: Roxbury CSA. *Roxbury Community Agriculture News*. 6(1).

Roxbury Farm Letter. 2001. June Newsletter.

Roxbury Farm Letter. 2007. August 6.

Roxbury Farm Letter. 2008. June 9.

Roxbury Farm Letter. 2009. June 15.

Scheib, F. 1994. *Capital District Community Agriculture: 1992-1994*. Unpublished historical records of early Roxbury CSA formation.

Chapter 6

THE FINANCIAL HISTORY OF ROXBURY FARM: 2000 - PRESENT[1]

ABSTRACT

As the operations of Roxbury Farm CSA have evolved over time, the finances of the farm have changed as well. This chapter presents the revenue sources of the CSA, detailing the share prices and how those are determined. Roxbury Farm CSA has, at various stages, offered different types of food shares – produce only, produce and fruit, and then the farm has started offering meat (pork, lamb, and beef) to members. This diversification has been important for the farm as the CSA no longer has to rely solely on produce, reducing the risk to the farm. In addition to the revenue stream of the CSA, the operating costs are presented to show what the farm spends money on. We will show that over time the farm has experienced efficiencies from use of technology and from learning over time.

6.1. INTRODUCTION

As Roxbury Farm has grown, additional revenue sources were developed, while at the same time implementing technologies that are labor and cost saving. On the revenue side, the farm added a meat share, a winter root share, an optional fruit share in cooperation with a local farmer, and has sold hay and

---

[1] NOTE: The budget data used and referenced in this book was obtained with permission from Roxbury Farm .

the farmer has done some consulting. The farm budget has evolved over time reflecting the overriding mission of the farm to enhance the life of the soil, bringing biodiversity to the land, providing a living for the farmers and income for farm labor, and training for future biodynamic farmers. The budget is a simple two-dimensional mirror for the way that resources and farm output are allocated. Not all value-added to the farm is recorded in revenue nor are all labor expenditures as expenses.

However, we can see some important developments over time indicating the financial health and sustainability of the Roxbury Farm CSA. We will now take a closer look at shares types and revenue over the first decade of the 2000s.

6.2. THE BASIC SHARE PRICE

At Roxbury Farm, the member share price configuration consists of three elements: first, the previous year's annual farm costs of production, including all material and labor costs; second, the delivery costs; and third, the distribution and administrative costs at the different sites. The setting of the share price historically has been the result of a conversation between the farmer and core members from each major cluster of members. More recently, the share price was set based on costs and feedback from the annual survey of members. Guidelines for the base share price generally include a mutual understanding of what the farmer needs to advance the mutually agreed mission of the farm, what is affordable for the members, the needs of apprentices, farm workers, and the farmers. Delivery and administration costs are added to the basic price in order to calculate the share price for each delivery site.

The share price is also influenced by wholesale and retail market prices. They provide a guideline to the high and low limits in setting the share price. This price range is set at a level above the price that the farmer could receive by selling the produce in the wholesale market but below the cost to the members of having to buy the same organic produce in the retail market. Historically, the Roxbury share price has consistently been in this price range. Since the share price is paid either in a lump sum or in a few installments in advance of the growing season, the farmer's operational costs and some capital costs can be done on a cash basis. Technically, on the installment plan, two-thirds of the share price must be paid before the delivery season begins and the remaining balance shortly thereafter.

Up to the time of the farm transition, the farmer was able to plan each year in conjunction with a group of core members volunteering to create a budget consistent with a production plan to match the number of member slots. An advantage of this budget method is that the farmer was not subject to the vagaries of the spot market for vegetables nor did he have to borrow money to meet operational costs or delivery costs. The members benefitted from having a predictable stream of fresh produce for the growing season and some input into the process of share price determination. After the transition, the farmers collected members' preferences about share price from the annual surveys and from less formal feedback from members. The farmers have reconvened meetings periodically with core members to discuss the farm, the farm budget, and other issues. These meetings have helped to restore some of the 'community' in CSA, which was missing in the years prior.

The share price from the year 2000 to 2010 in nominal (actual dollar amount) and real dollars (inflation adjusted), as well as the annual change in real dollars is shown in Table 6.1. The data shown is for the Kinderhook farm location. The base share price for all members is the production cost at the farm excluding delivery and on-site administrative cost. Those members who pick up their share at the farm have the lowest share price, while the Capital District and New York City share prices are higher and highest respectively because of delivery and administrative costs.

Table 6.1. Share Price and Membership Data for Roxbury Farm CSA: 2000-2010

Year	Share Price in Nominal Dollars	Real Share Price in Year 2000 Dollars	Change in Real Share Price (in %)	Planned Members	Actual Members	Excess Supply (ES) or Demand (ED)
2000	308	308	—	650	585	ES
2001	323	314	1.9	650	640	ES
2002	361	346	10.2	675	675	—
2003	375	351	1.4	675	675	—
2004	386	352	0.3	780	780	ED
2005	395	348	-1.1	925	925	ED
2006	404	345	-0.9	970	970	ED
2007	430	357	4.5	995	995	ED
2008	447	354	0.0	1040	1040	ED
2009	488	391	9.5	1065	1065	ED
2010	502	396	1.2	1100	1100	ED

Over the last ten years, the inflation adjusted real share price has risen 29%, or on the average of 2.9% per year although two years, 2002 and 2009, account for most of the rise. The reason for the 10.2% rise in 2002 shown in Table 6.1 was due to the heavy costs of transition from the old location, the cost of rented and leased land, the cost of hiring a farmer, the cost of fund-raising for the Capital Campaign, and the loss of some members in the transition. The farmer lost money during those two years not only because of the transition costs, but also because the farm was undersubscribed. From 2002 to 2009, the share price increased at a low but steady rate primarily because of the increased efficiency of farm help and experienced apprentices, the implementation of labor-saving farm equipment, and a closer match of resources to increases in the scale of operation. The oil price spike occurred in 2008, the year in which several members volunteered to send in a voluntary supplement to the existing share price. In 2009, the price hike of 9.5% needed to be built into the budgeted share price to make up sharp unexpected increases in energy and energy related cost increases especially since there had not been an increase in the previous year. Moreover, a substantial share price increase in 2009 could also be explained by capital improvements in the washing barn, packaging of produce for delivery, and the transition to biodegradable materials rather than plastic for weed control.

Though real inflation adjusted share price has increased over the first decade of the 21st century, the totality of these increases also advanced the vision of what the farm should be for both the farmers and the members. First, improvements in the washing facilities contributed to cleaner produce, more member satisfaction, and reduced food preparation times. Second, the pre-packaging of some produce and the fruit share contributed to facilitating pickup efficiency and convenience. And finally, the use of biodegradable "plastic" for weed control reduces the amount of toxic material at landfills, reduces labor costs, and contributes to a more healthy environment overall. In sum, the increased cost to members has an increased benefit counterpart for members, the environment, and increased work hours for farm help.

As indicated in Table 6.1, from 2004 to the present, there has been excess demand (ED) for shares. In conventional terms, this would place an upward pressure on the share price. The excess demand can best be explained by improvements in the quality of the share and convenience of pickup. These improvements have been beneficial to both the farmers and the members. The mutually determined share price takes into account the income needs of the farmers and farm expenses and shared risk of a shortfall in production. For instance, after a hail storm in early 2008 season, the farmers responded to members' notes of encouragement:

> "We greatly appreciate the notes of support and concern we received last week. Your support during the next few weeks means that we will come through the storm and be just fine. We don't have to worry about not being able to pay the bills or finding it difficult to pay our farm crew. By sharing in the risk of farming, you ensure that the farm will survive a difficult season and be able to begin a new one next year. And when we have a bumper crop, you share in the rewards. While we don't like disappointing you with small shares, we know that you are right there with us, and that keeps us going (literally and figuratively). So, we will all have to be patient together for the next few weeks while the damaged crops recover and while we wait for the tomatoes and sweet corn." (Roxbury Farm Letter, 2008)

One of the mutually determined financial goals of the farm is to provide benefits to Roxbury farm-workers of both a living wage and health insurance. With the support of members, this goal is achievable. However, this will require an incremental increase in the real share price over time.

Figure 6.1 shows the trend in income received from the basic produce share as a percentage of total farm income over the ten year period, 2000-2009. Note that the trend is a widening gap between vegetable shares and total farm income.

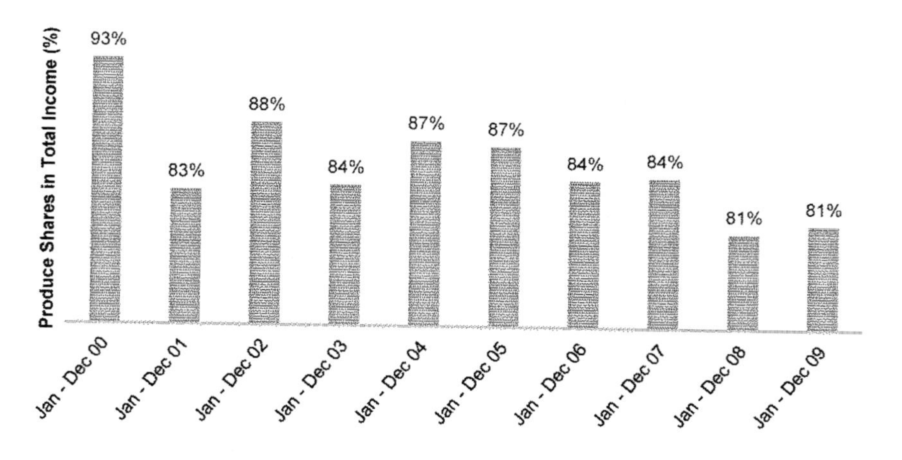

Figure 6.1. Income from Produce Shares as a Percentage of Total Farm Income: 2000-2009.

The decrease over time is explained by the increasing importance of three additional add-on shares available to members: the fruit share, the meat share and the winter root vegetable share. The fruit share is an opportunity for members to buy local non-organic fruit produced on a nearby farm. The meat share is an additional option resulting from the introduction of beef, lamb, and pigs to the farm. The winter share consists of a 40 pound box of storage vegetables consisting of potatoes, beets, onions, cabbage, and more. This optional share is delivered as an option after the ordinary 25-week delivery season has ended.

6.3. Fruit Share

In 2001, Roxbury offered a fruit share for the first time in collaboration with a local fruit farmer. The fruit was low-spray (integrated pest management) and was offered as an optional share to complement the vegetable share (Figure 6.2). The CSA farmers would purchase the fruit in bulk using the income derived from the members who signed up for and paid for the fruit share in advance. Members were informed of the fact that the fruit was not organic but local. There are no organic fruit growers in the Hudson River Valley that could meet the needs of Roxbury members.

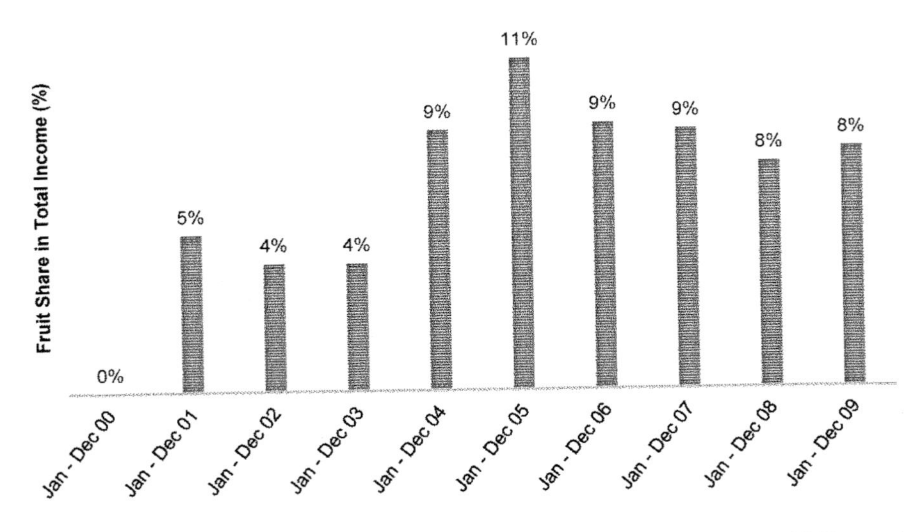

Figure 6.2. Income from Fruit Shares as a Percentage of Total Farm Income: 2000-2009.

The fruit share consists of many different varieties of locally produced fruit including strawberries, cherries, peaches, plums, and apples. Over the first few years, there were numerous member complaints about the quality and the lack of variety of the fruit. The member feedback prompted the Roxbury farmer to engage a different fruit farmer. The slow start is depicted in Figure 6.2 as the fruit share provided between 4% and 5% of farm income for the first few years. However, after the farmer established a new relationship with a reliable fruit farmer, the fruit shares became more popular among members. The fruit share as a percentage of farm income approximately doubled starting in 2004, and has leveled off at about 9 percent of farm income. However, since the membership has increased as well as the number of fruit shares, the amount of farm labor dedicated to packaging the share has increased.

There are several benefits to outsourcing the fruit. First, a local fruit farmer benefits by having a steady and predictable buyer throughout the growing season. On a couple occasions, Roxbury was the only local buyer for hail-damaged fruit because the CSA members accepted blemished but still good fruit. They realized that blemishes affected neither the taste nor the nutritional quality of the fruit. Ironically, and to the benefit of members, once the fruit is blemished by hail, the low-spray fruit becomes no-spray because the farmer ceases all additional spraying for economic reasons. Secondly, the CSA members benefit by being able to choose a CSA share with or without the fruit option. Third, the CSA farmers benefit through a slight mark-up to cover the wholesale cost of the fruit and the cost of packaging the fruit for the members. The packaging provides steadier employment for the hourly workers at the farm. The downside of the arrangement, of course, is that the fruit is neither organic nor biodynamic, a turn-off for some members. Since the fruit is fresh and delivered to the pick-up site at the same time as the regular share and the price is below the retail market price, members who participate in the fruit share find this a convenient way to purchase locally grown fruit.

6.4. Pork, Lamb and Beef Shares

Beginning in 2005, the farm began to raise free-range organic pigs, beef cattle and sheep in order to offer pork, lamb and beef shares in 2006. These meat shares were offered on an availability basis in addition to the vegetable share. The share is not a weekly or monthly share, rather an optional purchase available to members. After the animals are slaughtered and packaged, members are provided a list of cuts that are available along with the price per pound. Members then choose the amount that they intend to purchase. The meat is delivered frozen in coolers along with the delivery of the member's regular produce share.

The lamb and beef are grass-fed and free-range, making use of the parts of the farm not as well suited to growing vegetables. The pigs are fed some grain, vegetable culls from the vegetable operation, and forage for themselves on the wild roots from a fenced-in wooded area. The sheep and beef cattle are moved from pasture to pasture on a regular basis to graze. The cost of raising the animals is much higher per pound because of the cost of supplementary feed and processing. While the animals are a necessary component of a biodynamic diverse farm because they return nutrients to the land, the economic viability of the animal operation will depend on the willingness of the members to purchase the meat as it becomes available.

From the farmers' perspective:

> "The nice advantage of sheep and steers over pigs is that they make us completely self-sustainable in our food supply; they eat grass, which especially after all this rain has been in plentiful supply. For the pigs, we used to buy in pig feed from a local grain dealer that purchases his grain from local farmers. That has changed; this year he couldn't guarantee that the grains he purchased were GMO-free. As a result, we have secured another source of pig feed from Green Mountain Organics that will guarantee the feed to be free from any GMOs. Unfortunately, the cost of certified organic feed is much higher - the cost per ton went from $310 to $590. So, my dear pork lovers, you will notice an increase in the cost of pork this fall. If that means that we have priced ourselves out of the market due to our set of high standards on both animal ethics and feed source, we will be happy to give the forest back to the gnomes." (Roxbury Farm Letter, 2009)

However, as shown in Figure 6.3, despite the ad hoc availability of meat shares, income from meat sales has become 5% of the farm's income. The meat prices are much higher than retail non-organic meat, especially the better cuts. Starting in 2011, the farm offered meat shares at fixed prices. The plan consists of ninety pounds of meat during the season, much the same as the produce share, with regular delivery at two month intervals. Members will choose the type of meat but will receive a variety of cuts. The meat shares provide the opportunity for the farm and members to contribute to the sustainability of the farm by strengthening the biodiversity of the farm, increasing farm income, providing more work for farmer-workers and farm-related work in meat processing and delivery. In addition, the members benefit by having access to free-range animal products. The reconfiguration and timing of the meat share also allows the farmer to be pre-paid for the cost of raising the animals.

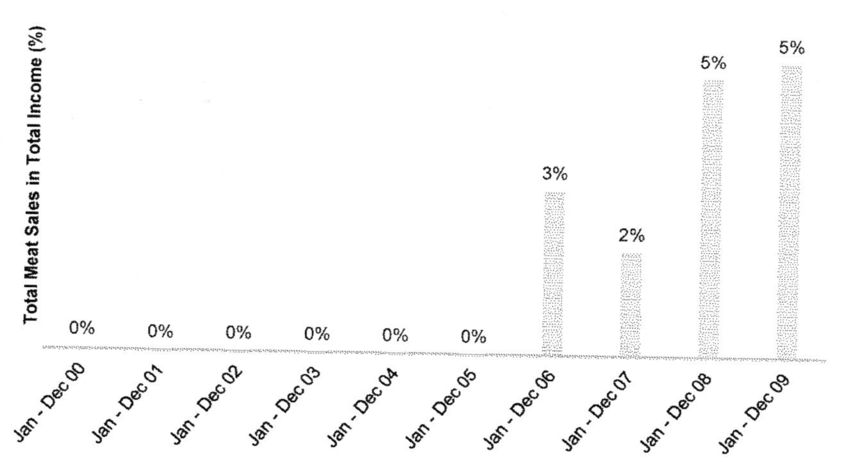

Figure 6.3. Income from Meat Shares as a Percentage of Total Farm Income: 2000-2009.

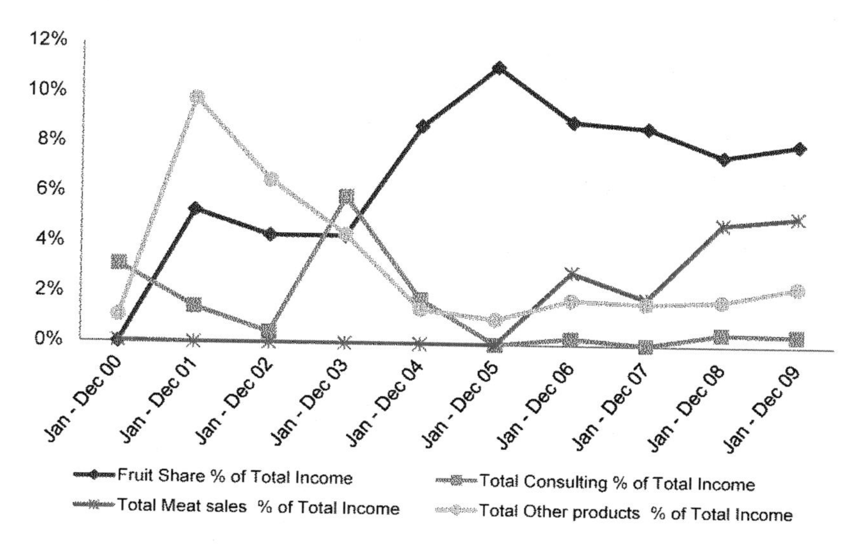

Figure 6.4. Non-produce Share Income for Roxbury Farm: 2000-2009.

6.5. Other Farm Income

The total income of the farm consists primarily of the basic vegetable share even though this element of income has declined over the past ten years and now provides a little over 80% of the farm revenue. However, a steadily growing membership has meant that vegetable production and acreage devoted to plants has increased. Most of the remainder of the farm income, as we have seen, has come from the introduction of the meat share, the fruit share, and the winter root share. The other sources of farm income are some hay sales, and very occasional consulting. A complete account of the minor components of total income are illustrated in Figure 6.4.

6.6. Long Term CSA Shares

The delivery truck breakdown, detailed in the previous chapter, offers an insight into the problems associated with economies of scale where one of the inputs into the process is inappropriate to the scale of operations. As the membership grew, the delivery truck, designed for light duty, short-range deliveries, adequate to handle a lighter load became less capable of the increased weight and more demanding delivery routes. As a consequence, the truck was subjected to excessive wear resulting in high repair bills. The light weight truck was one of the weakest links in the constellation of resources needed to run the farm operation efficiently. The truck capacity also placed an upper limit on membership that could be served in the delivery areas. Clearly, a new refrigerated truck with more capacity would allow for a slight increase in membership and more reliable delivery of fresh member shares. However, increasing membership would not generate enough revenue to justify borrowing money to purchase a new truck. From the social and sustainable perspective, a level membership of about 1100 seemed the right scale given the amount of farmland, greenhouse capacity, washing and storage space, work load, but a new truck was required to make everything efficient.

The farm suggested a creative method for the members to provide upfront money to purchase the truck without increasing the share price or the farmer losing income. The farmers offered 2, 5, and 10-year memberships at multiples of the share price at that time to raise enough money to buy a $50,000 truck. In this way, the farm would internalize the cost of the truck over time without having to raise the share price and the farmer benefitted greatly. The capital cost of the truck was covered interest free without having to borrow. The participating member benefits were a stable share price paid for in present dollars without having to reapply for membership until the contract ran out. The farm benefitted by gaining the efficiency of better matching the delivery truck to the optimal capacity with all the other inputs and to the farm operation itself. Furthermore, the farmer benefitted by not having to reduce his income by having to make payments on a new delivery truck. Moreover, this arrangement is an excellent example of how the risk of farm operation is completely shifted from the farmer to the consumer. Hence, this deal moved the farm closer to an appropriate economy of scale without impacting the farm's costs.

6.7. The Cost Side of Farm Operation

The data in Figure 6.5 gives an indication of the trends in farm costs over the last ten years measured as a percentage of total expense. The figure allows for an easy comparison of the average expenses for two five-year time periods, 2000-2004 and 2005-2009. The variation shows a significant change reflecting the farm goals of biodynamic diversity and good management in taking advantage of newer, improved technologies.

Increasing expenditures for the farm include overhead for animals, plant and soil amendments, plants and seeds. These increases show the farmers' priority in adhering to the biodynamic ideal of building up the fertility of the land and developing a balance of animal, plant, and human life on the farm. Bringing animals to the farm was undertaken to enhance the fertility of the land. In sum, these budget priorities could be considered an investment into the biodiversity and sustainability of the farm in the long run.

Areas of declining costs as a percentage of expenditure include office, insurance, interest, phone, postage, printing, supplies, and professional fees. There are two main areas in which cost savings have transpired over the 10 year period: one time transition costs and ongoing economies available by taking advantage of technological change. The transition costs in the early 2000s included much of the initial high cost of running the capital campaign to raise money for farmland as well as the high-cost printed material associated with running the campaign. The legal fees were also one-time expenses associated with the land acquisition, lease arrangements, and other transition issues.

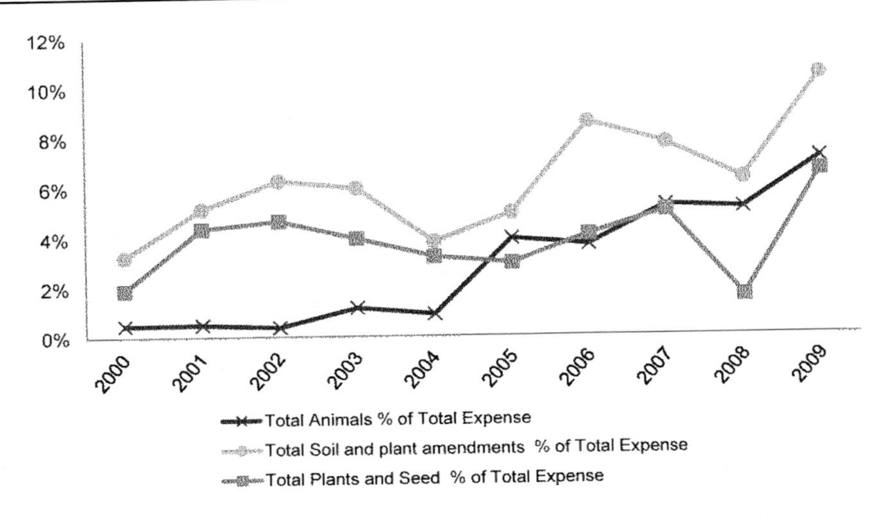

Figure 6.5. Three Largest Percentage Increases in Expenditures at Roxbury Farm: 2000-2009.

Figure 6.6 illustrates eight areas of spending decline over the past 10 years. Note the efficiencies that have become very pronounced in the past five years. The other major explanation of these relative cost savings is the use of computer technology for maintaining office records and the use of the internet to distribute over a thousand copies of the farm letter every week at a fraction of the cost of printed material. Roxbury Farm has benefitted from advances in communications technology thereby reducing phone expenses. Cell phones used by the farm crew enable a very rapid response to problems as they occur. Quickly solving problems saves both time and money. Even simple communication between farm crews and the farmer saves time that otherwise might have been wasted or misallocated.

Also, the long term CSA shares and declining interest rates, in general, have reduced interest expenses. The efficiencies of all the factors mentioned above have led to a reduction in administrative office expenses overall.

Another factor that does not show is the cost saving resulting from the members who volunteer beyond their work requirement at the delivery sites. The direct savings to the farm are considerable taking into account that the members have basically adopted the garlic crop from start-to-finish: garlic splitting, planting, and harvesting. Moreover, the members are involved in some weeding, and the harvest of three or more vegetables harvested on member work days in the fall. In this way, the monetary costs are held down thereby keeping down the potential share price increases.

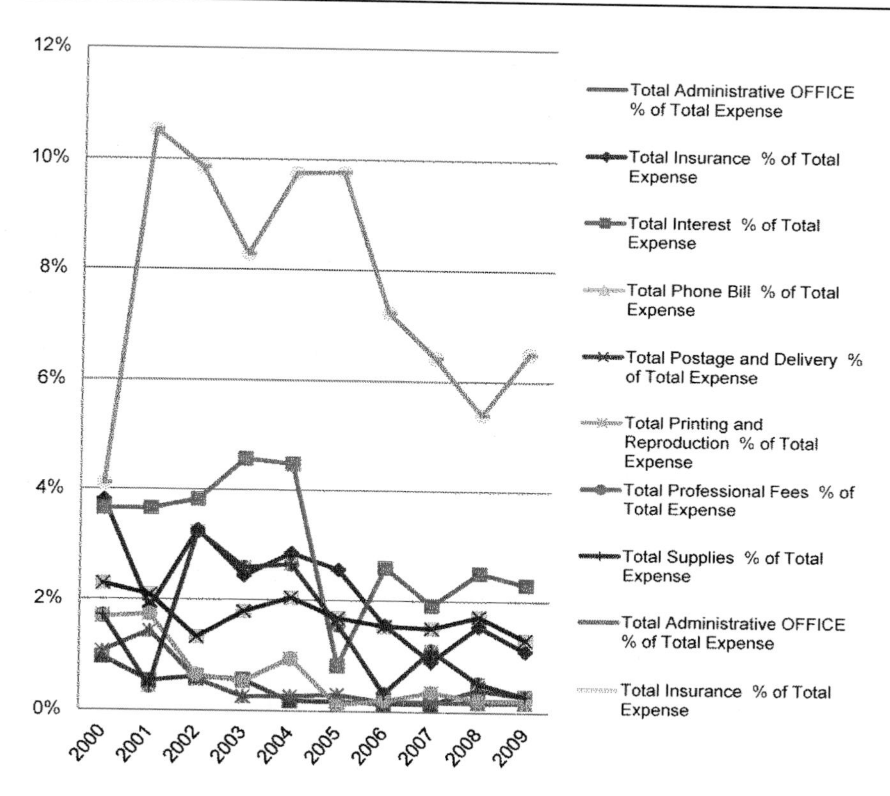

Figure 6.6. Eight Largest Percentage Decreases in Expenditures: 2000-2009.

CONCLUSION

This chapter has presented detailed financial information for Roxbury Farm CSA. Understanding the finances of the farm is important in order to see how the CSA operates from year to year. This relationship between finances and operations begins with deciding the member share price. Therefore, of vital importance is keeping good accounting records so the farmer can make a good estimate on what future costs will be so that the member share price can be set at an appropriate level to cover income for himself and the farm staff, operation costs such as seeds and fuel, as well as other costs associated with running the farm.

As was shown in Table 6.1, share prices for Roxbury Farm CSA have increased over time. However, the share price increases have not been uniform, as some years the price increase has been much more than others, corresponding with the challenges and financial issues described in Chapter 5. Of particular interest is the diversification of the farm over time, as evidenced by the decrease in produce shares as a percentage of total income illustrated in Figure 6.1.

In order to diversify the farm and increase total revenue for Roxbury Farm CSA, the farmer introduced fruit, pork, lamb, and beef shares that members could purchase in addition to their regular produce share. Furthermore, the farmer expanded revenue possibilities by offering a winter share. As the farm grew in reputation, the Roxbury farmer also had opportunities for consulting.

Meanwhile, on the cost side, the farm has experienced considerable efficiencies over time, largely due to better use of technology and learning over time. As the farm has explored more opportunities for generating revenue, costs have increased correspondingly. In particular, there has been a marked increase in costs for animals, plants and seed, and soil and plant amendments.

No history of a farm operation would be complete without also examining the financial conditions that correspond to the changes. This chapter has provided a thorough analysis of the revenue and cost streams of a CSA. The information provided is important because it shows the costs and revenues in relation to the scale of the farm operation, thus providing those interested in CSA knowledge about what to expect financially.

The next chapter builds the case study by presenting data from member surveys about the farm.

REFERENCES

Roxbury Farm Letter. 2008. July.
Roxbury Farm Letter. 2009. July 6.
Roxbury Farm Budget Data. 2000-2009.

Chapter 7

ROXBURY FARM CSA ANNUAL SURVEY RESULTS: 2003-2009

ABSTRACT

This chapter presents the data from a series of surveys, conducted for Roxbury Farm CSA members, from 2003 to 2009. Socio-economic and demographic information was collected and the surveys also asked questions on why people joined a CSA and their expectations. Additionally, members were asked if a variety of concerns, such as environmental, health, and financial, impacted their decision to join and retain membership in the CSA. Moreover, information on lifestyle and lifestyle changes from joining the CSA were asked as well. To the best of our knowledge, this dataset is the longest series for a CSA. Since members in CSA farms can be assumed to be homogeneous, the results presented provide considerable insight into CSA operation in general.

7.1. ROXBURY FARM ANNUAL SURVEY

Since 2003, Roxbury Farm CSA members have been surveyed annually on a variety of topics related to CSAs and specific questions related to Roxbury Farm. These surveys were conducted at the end of the season for members to provide their feedback to the farm about the way the CSA operated during that year.

Some of the information from these surveys has been used by the farmers to help them prepare for the following growing season, such as whether members plan on rejoining the next season, if members' expectations were met, what types of produce they would like to have more or less of, and what they think about the share price. Other survey questions were used to understand why the members joined the farm. Additional questions provided an insight into the ways members perceive certain social issues, such as those related to the environment and human health.

This chapter presents data from select survey questions and discusses how members perceive Roxbury Farm CSA and if these perceptions match the CSA theoretical concept.

7.2. Member Expectations and Membership Renewal

CSA memberships are driven by consumer demand. Therefore, in order to maintain a stable membership base, at least in the initial years of an individual's membership, the farmer must satisfy members' expectations in order for them to want to rejoin the farm for the following season.

As it will be shown later in this chapter, initial years of membership are particularly important for membership renewal because, after a certain period, members adapt to the stream of in-season produce, developing familiarity with new vegetables, and learning the details about the operation of the farm and the associated benefits and risks. Hence, variations in farm output become less important for the renewal decision over time.

Figures 7.1 and 7.2 show, for the period 2003 to 2009, the relationship between member expectations and membership renewal for Roxbury Farm CSA for the respondents of the questionnaire. Together, Figures 7.1 and 7.2 show a connection between membership renewal and the degree to which member expectations have been met.

However, if one were to sum the percentage of members who state that their expectations were either met or exceeded, he or she would notice that the percentage of membership renewals are higher. The likely reason is the same offered previously, that after a number of years of continuous membership, people learn about the CSA and the benefits associated with their membership. Due to this knowledge, these members understand the reasons for or are more willing to overlook the output variations that can occur year-to-year due to weather patterns or some unforeseen disruption in production.

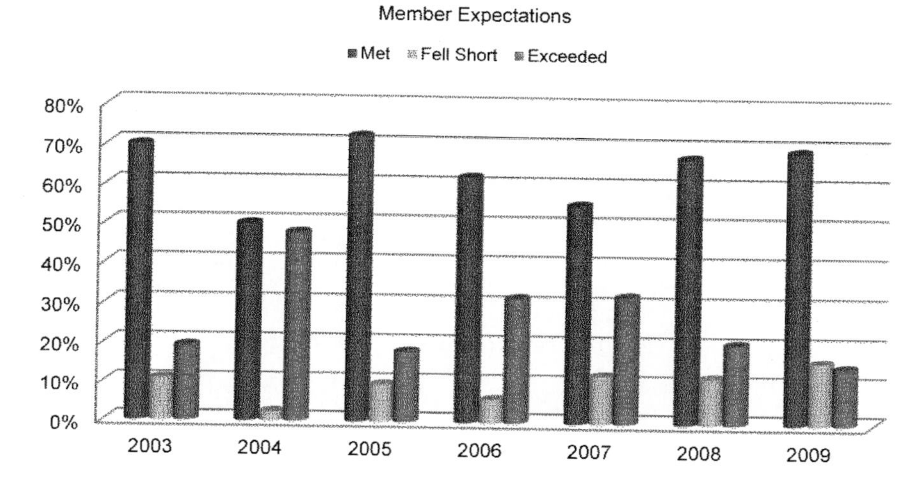

Figure 7.1. Percentage of Member Expectations that were Exceeded, Met, and Fell Short.

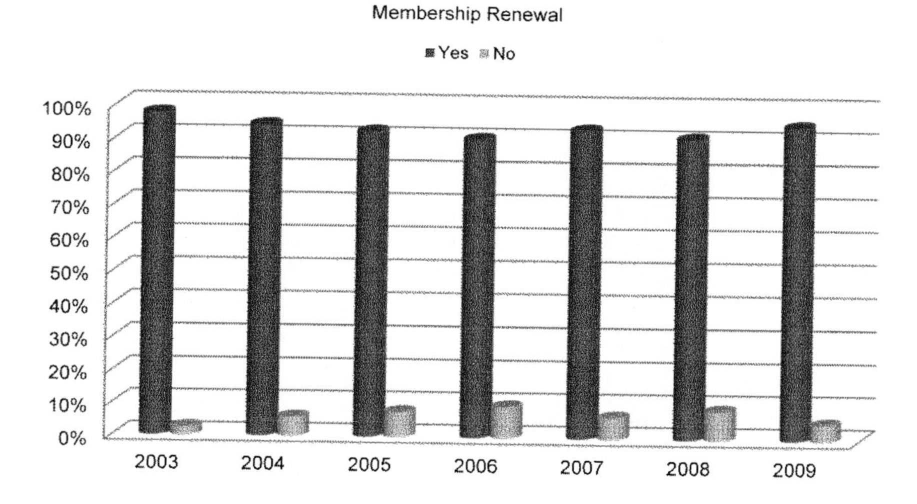

Figure 7.2. Percentage of Members That Will Renew Their Membership the Following Season.

Members were asked open-ended questions that provide some qualitative feedback to membership renewal. These responses highlight the notion that the farmer/member relationship is more than about market price. The nonmonetary factors most mentioned in rejoining the CSA are “community”, “support”, “support biodiversity”, “relationship.” While both price and quality are important, there is an explicit expression of desire to support the farm and the farmer. The following are a few representative responses to the open question “Why did you rejoin Roxbury Farm CSA?”:

- “The sense of community and knowledge of the food source, the commitment of Roxbury to the greater society.” Some members have a strong desire to be in association with a group of people who identify with the place their food is grown. However, they do not necessarily want to form a community in the traditional sense of the term.
- “To eat fresh, organic and support local farms.” Fresh and organic show up much stronger in the surveys than any other items.
- “The quality of the produce is great. I believe in eating locally grown food and supporting CSAs.” Quality is very important to members, even to the point that supporting local CSAs would fade if the quality was missing.
- “I wanted to support local biodiverse agriculture and to feed my family healthier, tastier food.” This comment shows the sentiment that while both biodiversity and healthier food are of concern for members, health issues are more important than biodiversity.

7.3. Share Price Dynamics: The Willingness to Pay More

As with any product, Roxbury CSA membership is also partly a function of the price of the CSA share. Like any business, the farm has operating costs that have tended to increase every year. As mentioned in the previous two chapters, the recent increases in the membership fee to cover the cost increases are somewhat associated with moving closer to the biodynamic ideal option since increasing the number of memberships they offer is not an option since shares are now near a maximum for the carrying capacity of the farm. However, the CSA farmer can only increase the share price as much as members are willing to pay. Therefore, the additional amount members are willing-to-pay provides the farmer with important information about the acceptable membership fee. Furthermore, the willingness-to-pay provides a measure of how much, in monetary terms, members value the CSA and all the benefits thereof. Since 2006 members at Roxbury Farm CSA have been asked how much more they would be willing-to-pay for their membership. The results are presented in Figure 7.3.

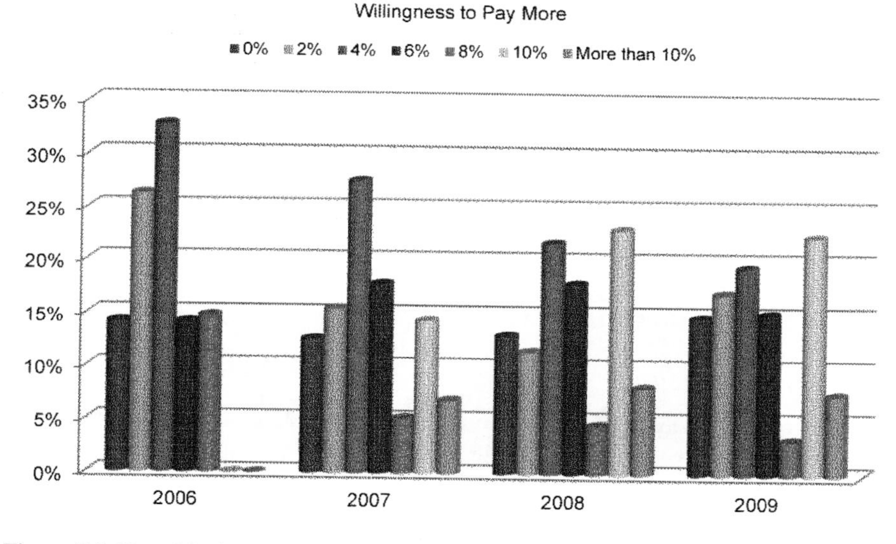

Figure 7.3. How Much More Members Are Willing-to-pay for Their Membership.

As Figure 7.3 illustrates, members were willing-to-pay, on average over the four years, approximately 5% more per year. However, the data also show some interesting patterns. From 2006 to 2008 there is a visible shift illustrating that members were willing-to-pay more for their membership, with many people willing-to-pay 6% or more the following year. Furthermore, the results also indicate that the majority of members not willing-to-pay any additional amount is highly correlated to the percentage of members not renewing their membership, a pattern suggesting that affordability is the primary issue for many members. In the recession of 2008-2009 and low inflationary period of 2009, the willingness-to-pay by members in the various percentage range cohorts shifted downwards from the previous years.

This lower willingness-to-pay is not surprising considering that the economic crisis caused many people to be unemployed or underemployed resulting in a decrease in their disposable income. In addition, the rate of inflation fell to near zero in 2009 so that any increase in share price was significant. The percentage of members willing-to-pay 0% to 2% more for their membership increased substantially while the percentage of people willing-to-pay more than 2% for their membership decreased. Yet, the decrease came primarily from the 4% to 8% range. This result is also likely to reflect some members' feeling that the farmer is taking advantage of excess demand; that membership fees have increased every year although members provided interest-free money for the purchase of the delivery truck and were asked by the farmer to pay a $25 supplemental fee during the oil price spike in 2008, but they saw no substantial increase in their share size. Interestingly the 10% and more than 10% categories only decreased slightly, indicating that the people willing-to-pay the most the following season have relatively inelastic demand while those willing-to-pay less than 10% more the following season have relatively elastic demand.

7.4. Motivation #1: Sustainable Agricultural Practices and Environmental Sustainability

In order to understand the factors that cause members to join the CSA, one must examine the perceptions of members about the agricultural practices used by the farm. Table 7.1 presents the importance of selected farm practices for members' decision to join the farm.

Table 7.1. Importance of Sustainable Farm Practices used on Roxbury Farm CSA in Members' Decisions to Join the CSA

	2004	2005	2006	2007	2008	2009
Soil Management	1.89	2.11	2.19	2.04	2.01	2.18
Crop Rotation	2.01	2.16	2.22	2.05	2.03	2.18
Cover Crops	2.13	2.3	2.34	2.21	2.2	2.34
Composting	1.87	1.96	2.08	1.9	1.99	2.06
No Pesticides Used	1.29	1.29	1.28	1.28	1.32	1.37
No Herbicides Used	1.31	1.32	1.31	1.3	1.34	1.38
No Groundwater Pollution	1.56	1.49	1.53	1.47	1.5	1.59
Healthier Habitat for Wildlife	1.73	1.77	1.83	1.69	1.68	1.82

1 = Very Important 5 = Not Important.

As one member commented on the 2008 survey, "I love that this is a biodynamic farm and that taking care of the land is top priority. This country needs to reacquaint itself with the small local farm and good growing practices."

Since the Roxbury Farm CSA uses biodynamic farm practices, all the methods members are asked about are contributing to sustainability. Perhaps the most important result is that each of these methods has lost importance for members over time. The most likely reason for this finding is that some specific farm methods used in production might not be too important to members. Rather, they might be more concerned with the broad farm practices used, mainly that the food is produced organically and may have begun to take these practices for granted.

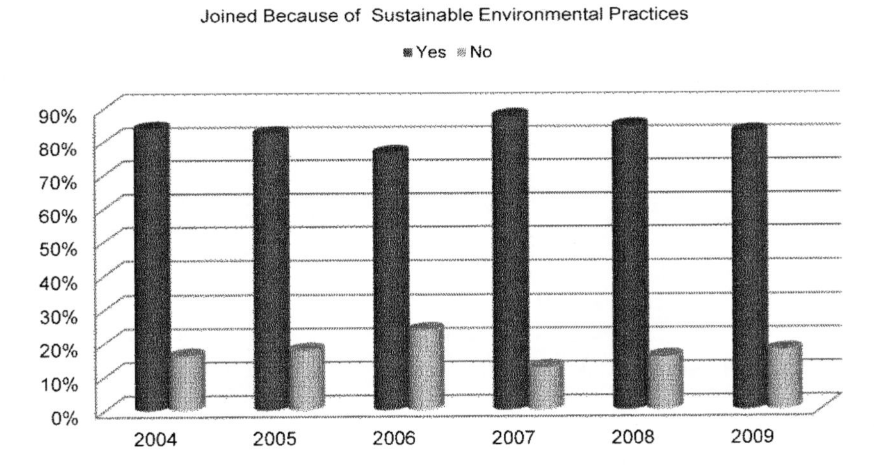

Figure 7.4. Percentage of Members Who Joined Roxbury Farm CSA because of the Sustainable Environmental Practices Used on the Farm.

The results of the survey confirm this conjecture as the most important farm practices members identify are that no pesticides are used. Members also seem to connect these farm practices to environmental health, as no groundwater pollution and a healthier habitat for wildlife are important as well. Figure 7.4 confirms this claim. As the figure shows, more than 80% of the members joined Roxbury Farm CSA due to the environmentally sustainable practices used on the farm.

7.5. Motivation #2: Production Methods and Human Health

Since members are making the connection between agricultural practices used on the farm and human health, they were asked about their personal health and the relationship between production methods and human health (Table 7.2).

Table 7.2. How Roxbury Farm CSA Members Describe Their Personal Health in 2009

	2009
Excellent	63.2%
Good	34.1%
Fair	2.3%
Poor	0.5%

The reason for these questions was to determine if members associated human health and environmental health. To obtain this information members have been asked a variety of questions over the years. In 2009, members were asked about their personal health and most members stated that they have either 'Excellent' or 'Good' health. This self-assessment is consistent with how members assess their personal weight situation, as shown in Table 7.3.

Table 7.3. Member Description of Their Personal Weight Situation in 2009

	2009
Very Overweight	1.8%
Somewhat Overweight	32.9%
About Right	63.9%
Somewhat Underweight	0.9%
Very Underweight	0.5%

Note that in a comparison of tables 7.2 and 7.3, the 63.2% reporting "Excellent" health corresponds favorably with the 63.9% reporting "About right" weight. "Good" health at 34.1% corresponds to "Somewhat

Overweight" at 32.9%, while "Fair" health at 2.3% corresponds to the 2.7% of "Very Overweight" or "Somewhat Underweight."

Therefore, that the members of Roxbury Farm CSA are health conscious should not be a surprise. Yet, when members were asked if they joined the CSA for health reasons, as shown in Figure 7.5, the percentage of members stating 'Yes' has declined substantially since 2004.

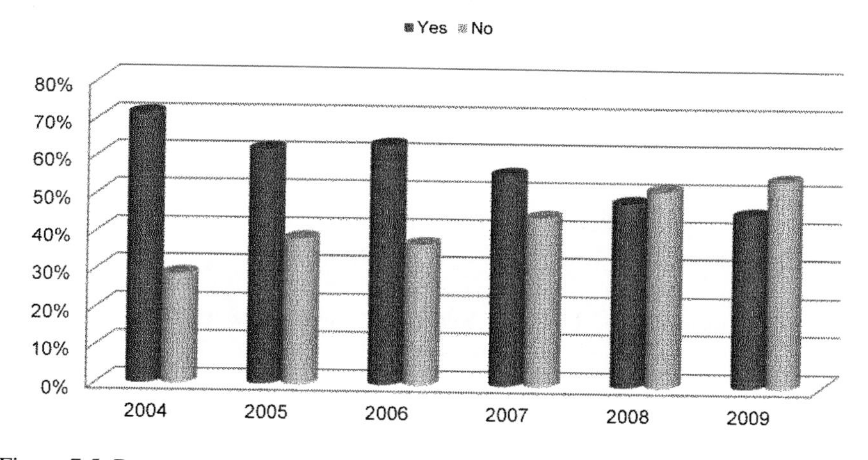

Figure 7.5. Percentage of Members That Joined Roxbury Farm CSA to Reduce Their Health Risks.

Of those who did join for health concerns, selected responses to the survey to the question "What health risks are you most concerned about?" suggests that many members are concerned about "health" in a sense that applies to children and to the world. Here are a few responses:

- "Reduce risk of pesticides on my daughter's health"
- "Cancer due to environmental toxins"
- "Risks from pesticides on both our personal health and everyone's health via environmental damage"
- "None in particular. We just feel that organic practices and eating locally are better for the earth and so better for everything on it, including us"
- "Poor nutritional value of industrially farmed produce, poor hygiene in processing industrially farmed produce and inadequate regulation of distribution".

One speculation about the results of Figure 7.5 is that members learn and adapt to the understanding of environmental and health issues over the years.

Since there is only one year of data for the individual health assessment of members, an explanation for the results illustrated in Figure 7.5 is difficult. However, the likely reason for the decreasing percentage of members joining the CSA to reduce health risks could be a result of the improving health conditions of the members. Additionally, since most members are likely to have an income greater than the population average, they would also be able to afford better healthcare to maintain good health. Therefore, over time the reduction of health risks for members became less important to them.

Adding credibility to this hypothesis are the results about the health-related reasons members provided for joining the CSA. While members strongly identify specific health-related reasons as their purpose in joining the CSA, as shown in Table 7.4, their perceived importance of health related reasons to join has decreased over time. This is consistent with the results reported in Figure 7.5.

Table 7.4. Importance of Health Reasons in Members' Decisions to Join the CSA

	2004	2005	2006	2007	2008	2009
No Contact with Pesticides	1.26	1.39	1.42	1.39	1.4	1.43
No Contact with Herbicides	1.29	1.4	1.41	1.41	1.41	1.44
More Nutrients in the Food	1.45	1.42	1.47	1.41	1.46	1.49
More Antioxidants in the Food	1.72	1.67	1.66	1.6	1.72	1.74
Less Toxins in the Food	1.32	1.37	1.38	1.31	1.36	1.4
Fresher Food	1.29	1.23	1.25	1.17	1.23	1.24
To Fight Disease	2.23	2.2	2.33	2.31	2.31	2.52
For Healthier Breastfeeding	3.14	3.09	3.26	3.05	3.3	3.32

1 = Very Important 5 = Not Important.

Therefore, the rationale for the decreasing importance of health-related factors for joining the CSA points to an interesting finding in these results. Members seemingly do not make the same connection between the agricultural practices of the farm affecting human health as they do with environmental health.

"No Contact with Pesticides" and "No Contact with Herbicides" has the same importance as "Less Toxins in Food" and only slightly less importance for members than for "Fresher Food." A seeming disconnect is the lower ranking importance of "More Nutrients in the Food" and "More Antioxidants in the Food" as these are directly related to health. One might conclude that members are more interested in avoiding toxins in the food than they are interested in increased amounts of substances that promote good health.

Another very interesting finding is the result "For Healthier Breastfeeding", which was asked only to breastfeeding women. As the table shows, not only has this health factor lost importance over time but healthier breastfeeding has consistently been the least important health factor over time. This result is very shocking as it suggests that breastfeeding women do not make the connection between what they are eating and the potential impact on the health and development of their baby. Why these results have occurred is unknown but the findings do suggest that health factors are a major reason for people to join the CSA and keep their membership active.

7.6. Analysis of Decision Factors to Join and Renew Membership

As presented in Tables 7.5 and 7.6, there are several other reasons for joining a CSA. First, unlike most of the other previously presented results, several of the reasons have increased in importance over time. For example, both "Economic Value" and "High Quality Produce" ranked higher in magnitude since the 2003 survey. It is not surprising that both these reasons improved in significance since organic food, in general, has become more popular and overall demand has increased, increasing the market price for organic food.

Ironically, the importance of obtaining organic produce, while highly rated at 4.53, has declined while the desire for high quality produce has increased to 4.87. This suggests that members care a little more about the appearance, freshness and taste than whether or not it is organic. This result could be because members are used to purchasing 'perfect' looking food at grocery stores.

Additionally, increasing demand has lead industrial agricultural firms, using old techniques without chemical pesticides, to mass produce organic vegetables in a similar manner to non-organic produce (i.e., same size, shapes, etc.). Because of the recent widespread availability of organic produce in supermarkets, members have come to expect high quality organic produce but also want the food at a relatively good price, particularly during the economic crisis of 2008 and 2009.

Table 7.5. Importance of Different Factors in Members' Decision to Join the CSA

	2003	2004	2005	2006	2007	2008	2009
Economic Value	3.36	3.73	3.76	3.8	3.78	3.92	3.87
Organic Produce	4.74	4.75	4.76	4.69	4.67	4.65	4.53
High Quality Produce	4.78	4.88	4.87	4.88	4.86	4.87	4.87
Community Building	3.64	3.78	3.82	3.73	3.72	3.87	3.52
Promoting a Healthy Environment	4.68	4.64	4.63	4.58	4.63	4.6	4.65
Support Local Farms	4.72	4.69	4.73	4.66	4.74	4.69	4.7
Share Risk with Farmers	3.9	4.01	3.96	3.83	3.93	3.94	3.86
Concern with my Health	4.46	4.38	4.29	4.36	4.28	4.32	4.11

1 = Not Important 5 = Very Important.

Table 7.6. Importance of Different Factors in Members' Decision to Retain Membership in the CSA

	2003	2004	2005	2006	2007	2008	2009
Economic Value	3.55	3.94	3.95	3.95	3.95	4.05	4.11
Organic Produce	4.72	4.83	4.81	4.74	4.76	4.71	4.61
High Quality Produce	4.79	4.88	4.9	4.88	4.9	4.9	4.89
Community Building	3.93	3.99	4.03	3.99	4.1	4.08	3.76
Promoting a Healthy Environment	4.70	4.74	4.74	4.77	4.8	4.77	4.75
Support Local Farms	4.75	4.88	4.85	4.81	4.91	4.85	4.80
Share Risk with Farmers	4.22	4.34	4.37	4.22	4.38	4.38	4.28
Concern with my Health	4.48	4.57	4.45	4.51	4.45	4.46	4.31

1 = Not Important 5 = Very Important.

As shown in Table 7.6, the same two factors, economic value and high quality produce, had increased importance for membership renewal decision. In addition, "Promoting a Healthy Environment", "Support Local Farms", and "Share Risk with Farmers" also increase in importance over time as a consumer decides to retain the membership.

The improvement in these categories after a consumer has been a member is not surprising since part of being a CSA member is learning about issues related to local agriculture and to the CSA experience. The "Promoting a Healthy Environment" reason is consistent with results presented earlier. The "Support Local Farms" and "Share Risk with Farmers" show that members learn about the importance of local agriculture and the impact agriculture has on rural economic development.

Community building has become less important for a number of reasons: new members are more interested in fresh organic produce; membership is more spread out with an increasing number of delivery sites; CSA administrative work has been centralized into the farm office; the farmer has not actively promoted the community aspect of the CSA; and there has been fewer crises that might motivate more community building.

7.7. ANALYSIS OF CONVENIENCE AND LIFESTYLE CHANGE

Since CSA members exhibit learning through their experience of membership, a natural progression would be to examine if and how have they changed their lifestyle. A CSA member does have to alter her/his lifestyle in one or more of the following ways: a week's supply of seasonal produce must be picked up on a certain day during a prescribed time period; some time must be spent cleaning and preparing the food; the members daily food preparation is now dictated by the in-season vegetables rather than menu-driven grocery shopping; vacation and restaurant meals make food planning even more complicated since the arrival of member shares arrive inexorably over the growing season; and members must contribute time working either on the farm or at the pick-up site, for some this is a major interruption of life-style. Therefore, there is a convenience factor in being a CSA member.

Table 7.7 presents a list of factors that consumers must contend with as a CSA member. As one might expect, several of these factors did not rate well. "Parking", "Hours", and "Day of Week" rated the lowest of the choices offered to members.

Table 7.7. Convenience Factors for a CSA Member

	2003	2004	2005	2006	2007	2008	2009
Hours	4.2	4.08	3.94	3.84	3.83	3.86	3.99
Location	4.18	4.31	4.27	4.22	4.14	4.17	4.29
Day of Week	4.16	4.1	3.99	3.97	3.96	3.93	4.07
Distribution Set-up	4.56	4.34	4.3	4.23	4.26	4.23	4.27
Parking	4.4	3.86	3.78	3.88	3.72	3.68	3.8

1 = Very Inconvenient 5 = Very Convenient.

"Parking" scored the lowest of all the factors most likely because many of the pick-up sites are at members' homes or other locations where off-street parking is limited. "Hours" and "Day of Week" are ranked low because both the time and day of pick-up are decided independently of members' input and result in inconvenience to some members. One explanation of the trend toward less convenience is that the pick-up schedule was initially determined jointly by the farm and new members as new sites were established. Hence, newer members are nearly powerless to change either the pickup day or times.

Table 7.8. Changes in Lifestyle Experienced by Roxbury Farm CSA Members

	2003	2004	2005	2006	2007	2008	2009
I am Eating More Vegetables in Season than before I was a CSA Member	92.0%	90.2%	86.2%	67.6%	60.7%	65.3%	67.9%
I am Cooking More than Before I was a CSA Member	43.7%	53.0%	54.7%	39.2%	45.4%	40.4%	41.5%
I am Planning More Meals than before I was a CSA Member	36.8%	43.9%	42.4%	37.2%	36.1%	35.2%	42.9%
I am More Conscious of the Effects of Food Production and Consumption on the Environment than before I was a CSA Member	64.4%	75.8%	71.7%	79.8%	84.7%	88.3%	71.9%
I Have Altered My Lifestyle in a Positive Way Now that I am a CSA Member	51.7%	49.2%	50.2%	41.4%	46.4%	43.1%	47.3%

Other reasons besides the convenience factors cause CSA members to alter their lifestyles. Table 7.8 presents the survey results of how members perceive some of these issues. Members were asked to identify all statements which they agree with, therefore the percentages add to more than 100%.

Consistent with the results presented previously, members have identified "I am More Conscious of the Effects of Food Production and Consumption on the Environment than before I was a CSA Member" as one of the most important lifestyle changes they have experienced. One interesting result is the major decrease (about 1/3) in the percentage of members choosing "I am Eating More Vegetables in Season than before I was a CSA Member".

The most likely reason for this result is that consumers have maintained a continuous membership so their consumption of vegetables would not be increasing over time since they have presumably been eating a lot of vegetables from the farm to begin with even though two-thirds are still reporting that they eat more vegetables in season. If new members are already eating more vegetables before they join, that would explain the drop from 92% in 2003 to 67% in 2009.

More than 40% of respondents indicated that they are "Cooking More" and "Planning More Meals" than before they were members of the CSA. This result is due to the volume of food that members receive as part of their share; there is no other choice but to plan and cook more meals and consume/freeze the product, throw it away, or give it away. As a result members change their eating behavior.

Table 7.9 confirms this claim illustrating the number of servings of fruits and vegetables per day that members eat before and after the season. Not surprisingly, members eat a large number of servings each day during the season. As expected, after the growing season is over many members continue eating a large number of servings of fruits and vegetables each day, though there is a shifting toward 3 rather than 4 servings a day.

Table 7.9. Number of Servings of Fruits and Vegetables per Day Members Eat Before and After the Season

	Before	After
0-1	0.0%	3.1%
2-3	25.9%	44.4%
4-5	50.0%	39.9%
More than 5	24.1%	12.6%

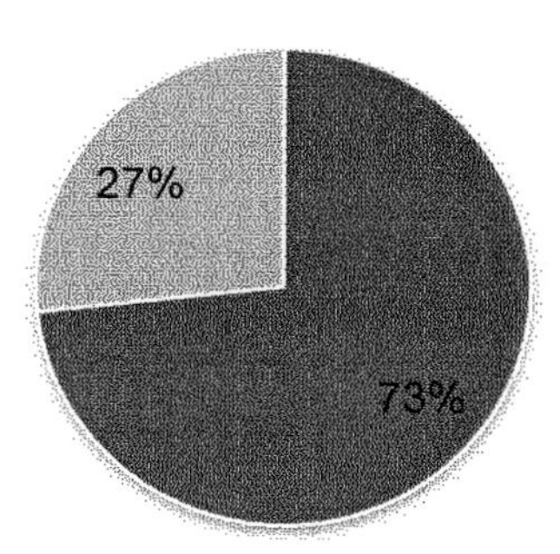

Figure 7.6. Percentage of Members That Purchase Organically Grown Food after the CSA Season in 2009.

Therefore, members are exhibiting some lifestyle changes during the season, while regressing toward prior consumption habits during the off season. Since members continue to eat several servings of fruits and vegetables each day even after the CSA season, they were asked if these fruits and vegetables were organically produced. This question has been asked only once, in 2009, so far. The responses are shown in Figure 7.6.

As the figure shows, approximately three-fourths of the members responding to the survey indicated that they continue to purchase and eat organically grown food after the CSA season is over though they will consume lesser amounts.

7.8. CSA'S IMPACT ON ALTRUISTIC BEHAVIOR

Since members are eating so many servings of fruits and vegetables and they state that they are planning and cooking more meals than before they were a CSA member, many members have difficulty keeping up with the flow of produce into their household as the season progresses. Hence, the next logical question was "How Much of Your Food Share Do You Give Away Each Week?" The results are presented in Figure 7.7.

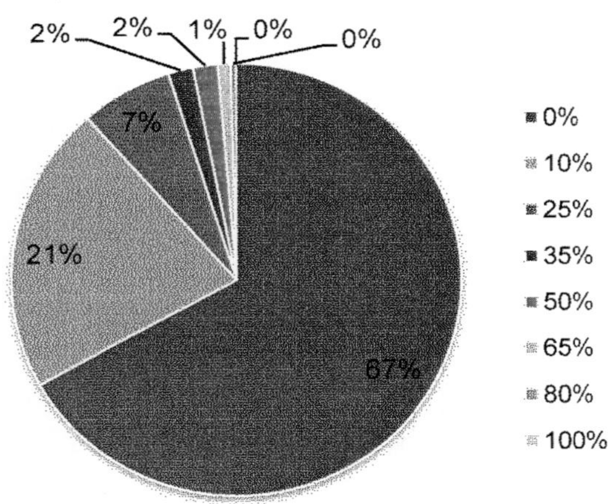

Figure 7.7. Percentage of Food Share Members Give Away Each Week, 2009.

As shown in the figure, the majority of members give away less than 10% of their food share each week. However, slightly more than 10% of members give away 25% or more of their food share. This result suggests that these members exhibit altruistic behavior. Many members are pleased to know that all the shares that are not picked up are delivered to a local food pantry or soup kitchen.

If members give their surplus food to friends, relatives or neighbors, then more potential members are introduced to the CSA grown produce, possibly explaining some of the excess demand for memberships.

7.9. CSA'S IMPACT ON CIVIC ENGAGEMENT

Because members are experiencing changes in behavior and lifestyle they were also asked a series of questions related to community and public policies to find out if there was a spillover effect of private life-style change into a behavioral change into the public arena. In 2009, Roxbury Farm CSA members were asked if being a member of the CSA caused them to increase their activity in civic involvement in a variety of public policies related to agriculture.

Table 7.10. CSA Membership Causing an Expansion of Civic Involvement in Various Policies, 2009

Energy Policy	3.8%
Healthcare Policy	8.6%
Environmental Policy	24.8%
Local Economic Development	20.0%
Agricultural Policy	42.9%

Table 7.10 shows that civic involvement has increased since any percentage greater than zero means that CSA membership has motivated an increase in involvement in a variety of public policies, most notably agricultural policy.

The other areas which rated highly for member involvement were environmental policy and local economic development. These results are not surprising as agricultural policy is directly related to the farm and environmental policy and local economic development are related. First, the production practices of the farm coincide with sustainability and embody sound environmental policy. Secondly, keeping money in the local economy by members choosing to purchase local products coincides with the "locavore" movement of consuming locally produced food products.

A surprisingly low percentage of members increased their involvement in energy or healthcare policy. Considering how important health is as a reason to join the CSA, an unusually low percentage of members, 8.6%, became civically involved, as it is for only 3.8% of members to be active in energy policy perhaps because the farm is not seen as providing examples of energy savings other than saving on delivery costs. However, given the disconnect that members have exhibited with other questions about healthcare; the low civic involvement in healthcare policy is consistent with the results for other survey questions. That members do not get involved with energy policy suggests that they do not understand the long-term connection between sustainable agriculture, industrial agriculture, and energy policies. Furthermore, on a larger scale, the result shows that members do not understand the connection between agriculture, the environment, and energy policy; in particular biofuels. In Anna Lappé's book *Diet for a Hot Planet,* she discusses the lack of public discussion about the contribution of industrial agriculture to greenhouse gas production. She notes (Lappé, 2010: 11) that

> "...including emissions from the production and distribution of farm chemicals from land as it's transformed to make way for crops and livestock and from energy for factory farms and food processing - and the entire global food chain may account for roughly one third of what's heating our planet."

However, all of the results suggest that there is a level of learning on members' part that entices many of them to become more involved in public policies that affect agriculture and that if further information would be provided on energy and health policies in relation to agriculture, members would become more civically involved. For some, joining a CSA is a political act, for others, membership prompts more civic engagement.

7.10. CSA's Impact on Community Building and Social Justice

In the 2009 survey, members were asked about other social issues related to CSA. As shown in Table 7.11 only one-quarter of the members think community building is realized in CSA, while more than one-third believed that it was not.

Table 7.11. How Members Perceive Community Building and Social Justice in CSA

	Is Community Building Realized in CSA?	Does CSA Improve Social Justice?
Yes	25.3%	31.1%
No	34.4%	10.0%
Do Not Know	40.3%	56.9%

This finding is a little surprising considering that one of the goals of CSA is to develop or foster community among members. Therefore, these results suggest that the majority of members socialize very little with each other; rather they collect their share each week and then leave the pick-up site. Thus, at least for Roxbury Farm CSA, community building is conceptual for at least 3 out of 4 members. This finding raises a larger question of what "community" means to the Roxbury CSA, especially since it is large compared to others. This result could be due to the lack of member input actively sought by the Roxbury farmer or what some members may perceive as his lack of willingness to respond to their needs, wants, or concerns. The farmer has this luxury as there has been excess demand as shown in Table 6.1.

Less than one-third of members responding to the survey think that social justice is improved by CSA. There is a surprisingly high percentage of members that think that social justice is improved by CSA considering that the membership fee for a CSA is a large amount of money that must be paid in full at the beginning of the year that seemingly prohibits lower income people from joining the CSA. Yet, despite this barrier to entry for membership, these members think that social justice is improved. The likely reason for this is that Roxbury Farm CSA provides extra food to each delivery site to be donated to a local food shelter at most of the delivery sites, such as the City Missions of Schenectady and Albany. Some members and sites subsidize the share price (scholarships) for low income members. The farm is willing to work out an extended payment plan for those persons who are economically distressed.

Furthermore, members may be linking the sustainable agricultural practices of the farm to social justice because low income people are, in general, more adversely affected by high levels of pollution caused by industrial agricultural practices. They may want to support the farm workers at Roxbury both financially and affirm that workers of Roxbury Farm are laboring in a safer environment than workers at conventional farms.

7.11. CSA's Impact on Food Security

These findings are consistent with the responses members provided about food security, a social justice issue in the United States. Members were first asked if CSA increases food security in the United States.

Table 7.12. Does CSA Increase Food Security in the United States?

	2005	2006	2007	2008	2009
Yes	75.6%	66.1%	78.0%	75.6%	74.9%
No	1.3%	1.4%	1.2%	0.5%	4.6%
Do Not Know	23.1%	32.5%	20.8%	23.9%	20.5%

As shown in Table 7.12 the majority of members that answered the survey think that CSA increases food security.

The most likely reason for this is that members are making the connection between local agriculture and food security, understanding that the monoculture prevalent in industrial agriculture is risky as the produce can be easily contaminated or destroyed by natural or manmade causes and the global trade of produce could easily be disrupted or halted. This hypothesis is validated by members' responses to whether they believe that food security is a problem for the United States; responses are provided in Table 7.13.

Table 7.13. Is Food Security a Problem for the United States?

	2006	2007	2008	2009
Yes	72.2%	78.7%	80.7%	83.2%
No	27.8%	21.3%	19.3%	16.8%

As illustrated by these results, the vast majority of members surveyed think that food security is a problem. Interestingly, over four years (2006-2009), the percentage of members who think that food security is a problem increases nearly 10%. This finding suggests that over time members have learned about food security issues in the United States and believe that CSA is a potential solution. This result makes sense since CSA can be thought of as a proxy for local agriculture; which is consistent with the results of the question on local agriculture presented earlier in Table 7.5. Furthermore, the increase in the percentage of members stating 'Yes' could also indicate that members sense that food security in the United States has gotten worse since 2006. For instance, fresh and processed fruit and vegetable imports have increased substantially over the past few decades in the United States and members are likely to have seen this increase by examining place of origin labels on their food. According to Food & Water Watch (2008) "the FDA (Food and Drug Administration) found that imported fruit is four times more likely to have illegal levels of pesticides and imported vegetables are twice as likely to have levels of pesticide residues as domestic fruits and vegetables."

CONCLUSION

The results of member surveys presented in this chapter provide empirical data to confirm the theoretical construct presented in Chapter 4, as well as provide context to the history of Roxbury Farm CSA presented in Chapters 5 and 6. Beyond providing substantiation for the previous chapters, the survey results presented in this chapter provide important information on how members that join Roxbury CSA think and perceive important public policies related to agriculture. This information can be used to support public policies aimed at expanding local, sustainable agriculture. While the results presented are from only one CSA farm, it is very likely that CSA members are a relatively homogenous group to some extent. Members of Roxbury Farm CSA tend to be well educated and earn a relatively high income. Little data exists on the socioeconomic status of CSA members but we assume that members in other CSAs have similar characteristics. The following chapter builds upon the previous chapters to examine some of the important issues related to agriculture today and seeks to explain how sustainable agriculture, especially CSA, can be used as a solution to these problems.

REFERENCES

Food & Water Watch. (2008). *The Poisoned Fruit of American Trade Policy: Produce Imports Overwhelm American Farmers and Consumers.* www.foodandwaterwatch.org

Lappé, A. (2010). *Diet For a Hot Planet: The Climate Crisis at the End of Your Fork and What You Can Do About It.* New York: Bloomsbury.

Chapter 8

Concluding Comments

Abstract

This chapter concludes the book with a discussion of our modern, industrial agricultural production and the negative aspects associated with it. As the environmental health impacts have already been largely discussed in the literature, we go into detail about the human health effects of industrial agriculture. Human health is impacted in a variety of ways, ranging from obesity to potentially cancer. These adverse health effects cost society and can be partially to completely averted with sustainable agriculture. Lastly, a brief argument of how sustainable agriculture can be used for economic development purposes is presented. This is potentially very important in developing countries as they could develop their agricultural sector to generate rural economic growth while at the same time improving their food security. Lastly, the book concludes with a few closing remarks.

8.1. Introduction

Agriculture, something many people today take for granted, is arguably the most important sector of the world economy. Until the dawn of modern, industrial agriculture, the vast majority of people understood the importance of the land because they worked the land for their food and occupations. Therefore, their connection to the land was intensely personal. They understood that ignoring the connection between agriculture and the care of the land could mean starvation. However, today's world population, particularly people in Western countries or urban centers, has forgotten the importance of this connection to the land. Energy-intensive and chemical applications in food production and storage, as well as transporting food long distances, have made food so readily available and easy to purchase to so many people that food is now often taken for granted. Ironically, over half of the world population goes to bed hungry every night. Adding to this disconnect is that Western governments have intervened to ensure a steady supply of inexpensively priced food available for their citizens by providing agricultural subsidies to producers. For example, Americans spend less than ten percent of their income on food (6.6% in 2012) and approximately thirty minutes a day on meal preparation (Wolfe, 2014; Hamrick et al., 2011). In fact, food has been in a deflationary state in comparison to other products as real food prices have decreased over the last several decades. Only recently has this trend reversed (Durden, 2014). Thus, one should not be surprised at this disconnect because people in the West simply do not have to think about where their food comes from, how it is produced, or the impact of the production method on the environment. Nor do they think about the true cost of their food.

8.2. Why Food Is not Cheap

Food is not 'cheap.' When the hidden costs to the environment, the health of the population, and agricultural costs are included, Western societies pay at least three times for their food produced by today's modern, industrial agriculture; once for the price at the store, a second time for taxes used to subsidize farmers to grow food for animals, people and fuel, and a third time for costs of environmental damage and public health side-effects (Pretty, 2005:51).

Changes in public-policy and technology have led to the corporatization of agriculture, which in turn has resulted in monocropping in the name of efficiency and higher profits. These changes, spurred-on by the availability of relatively cheap fossil fuels used to develop chemical inputs (fertilizers, antibiotics, hormones, and pesticides) and more fuel to run irrigation systems, have come at a great expense to the environment. The land at many of these industrial agriculture sites has been polluted to the point that it has lost the capacity to be productive without additional chemical applications because soil fertility has been destroyed. Furthermore, these external inputs to production have significantly damaged or destroyed the local groundwater and even distant ecosystems downstream. For example, fertilizer runoffs from the

Mississippi river valley have created a dead zone in the Gulf of Mexico "as big as the state of New Jersey—and still growing. By fertilizing the world, we alter the planet's composition of species and shrink its biodiversity." (Pollan, 2006: 47)

This approach to agriculture has affected more than environmental health; it has also had a major impact on public health which has gone largely unnoticed for many years until the 1980s when a string of highly-publicized outbreaks of *E. coli* (Armstrong et al., 1996) and mad cow disease (Cleeland, 2009) raised the issue. Additional attention continued throughout the 1990s with further outbreaks and additional problems such as *Salmonella* (Centers for Disease Control, 2013) and manure leaks at hog farms (Marks, 2001). As the connection to public health became clearer for the public, food journalism surged in popularity; various books, documentaries, and television shows became extremely popular, once again connecting the consciousness of much of the populace to food and agricultural concerns. More recently, there is a small but growing movement of "locavores", people who try to consume food grown close to home to save energy on long distance transport and to support local farmers. Media coverage has also helped to make people acutely aware of the health problems associated with industrial agriculture.

Moreover, the public has also become aware of the obesity epidemic that has stricken parts of the world for the first time in human history, especially more affluent nations, though obesity is emerging in developing countries too. Highly processed food and the ingredients in them, such as corn syrup, have come under increasing scrutiny for possible links to obesity, as well as cancer and diabetes, among other public health problems. Current public-policy towards agriculture in developed countries encourages the consumption of calorie-dense, nutrient-poor, industrially produced food, typically containing a variety of manmade potentially toxic additives, that causes people to accumulate and conserve energy as fat, as well as induce them to eat more of the types of foods that created the health problems in the first place (Wallinga, 2010; Brownwell, et al., 2009; Foresight, 2007). Certainly, one of the key developments of the obesity epidemic with the poor and middle-classes in developed countries is the proliferation of inexpensive fast-food options when compared to healthier, more nutritious food like fruits and vegetables. Those people and families on tight budgets often have to pass on purchasing fruits and vegetables due to the cost of these foods. Furthermore, these classes are often not well-educated on the nutritional benefits of food for their health.

Additionally, the manmade chemical inputs (fertilizers, antibiotics, hormones, and pesticides) used in producing food are believed, by many, to be carcinogenic. While there are no epidemiological studies that provide proof of direct causation between cancer and human consumption of these manmade chemicals in food, largely because of the myriad other outside influences that also contribute to the contraction of cancer. However, to those that believe in this correlation, the risk of cancer developing many years after consuming food grown on industrial farms is clear. The amount of circumstantial evidence and evidence from controlled animal studies is so substantial that many believe that there must be a connection. Moore (2002) explicitly describes the human health effects of exposure to these chemicals. External chemical inputs, such as fertilizers, are inhibitors of plant phenolics, or flavonoids which are effective in preventing cancer, heart disease, and age-related neurological disorders like Alzheimer's disease. Chemicals result in oxygen radicals that damage DNA, causing mutations that can lead to disease. Long-term oxidation of cell structures can result in nervous system disorders like Alzheimer's and Parkinson's disease (Cummins, 2003). In addition to these health concerns, milk from cows injected with hormones and antibiotics have long been suspected to cause early puberty in girls and possibly cancer.

The realization that 'cheap' food is not so 'cheap' once all the adverse effects are considered has helped to accelerate the sustainable food movement. At first, many thought this movement was a fad, but as time has passed sustainably produced food became more popular.

8.3. THE CROSSROADS OF AGRICULTURE

Although sustainable agriculture has become increasingly more desirable in the marketplace, the global agricultural system is at a crossroads. Modern, industrial agricultural farming methods have been so destructive that an ever-increasing amount of external inputs are needed to achieve the same amount of production because various organisms and insects have mutated thus diminishing the effect of chemical applications to defeat them. Clearly, modern, industrial agriculture is unsustainable over the long run both environmentally and economically. For example, in 1980 the United States yielded 15 to 20 tons of corn per application of a ton of chemical fertilizers. By 1997 the yield dropped to 5 to 10 tons of corn for the same ton of fertilizer (McKenney, 2002:128).

Ironically, only the few corporations which control most of the inputs to agricultural system profit from the system that has helped to create food insecurity for many people in the world (Farm Aid, 2013). However, the chemical, processing, and retail establishments that depend on industrial agriculture will not let the system go away easily. Oligopolistic power over farmers and consumers, as well as political power to influence subsidies and regulations will slow the transition to sustainability unless the demand for sustainable agriculture becomes so large that firms will see profit opportunities.

Affordable access to good, nutritious, and safe food is a challenge for all nations in the world today. Many countries engage in a policy of specializing in the production of certain crops, such as grain and corn, and then exporting these subsidized crops thousands of miles around the world. The concept is simple, trade opportunities will help to stimulate production. However, subsidies create an artificial comparative advantage, one that would otherwise be nonexistent. The resulting underpriced agricultural exports to world markets have created a dependency in many countries on imported food because their local farms, many of which were sustainable, were forced to cease operation because they could not compete with such low prices. The social costs of dislocated small farmers has been enormous. Comparative advantage does apply to some unique food products that can only be produced in local microclimates, but world trade in food is severely distorted because most of the trade is in food staples which are heavily subsidized, especially in North America and Europe.

Although there is a substantial amount of food produced, enough to feed the world's population, access to this food is limited because production is concentrated in relatively few regions. Geographical and environmental issues compound the dependency on food staples produced thousands of miles away from the countries that are importing them. Population increases over the next couple of decades and climate change, accelerated with as much as one-third of the greenhouse gases generated by industrial agriculture (Lappé, 2010), will only add to food (in)security issues, especially as local farms are forced out of business by cheap food produced by industrial agriculture. In the long run, these problems will increase the demand for food leading to an increase in prices which will disproportionately affect the poor, both farmers and consumers. Food price pressures will be exacerbated by the sustainable energy movement, in particular biofuels.

For example, during the petroleum price spikes in 2008, the demand for 'sustainable' energy, especially ethanol, made from agricultural products increased, which caused an increase in food prices since production of biofuels uses farmland that would otherwise be used for food production. Hence the price increase was both direct and indirect; the direct price increase was due to the increase in demand, as well as from shifting existing food production (mostly corn) to ethanol production. This production shift further deepened monocropping which caused the prices for other crops to increase as their supplies decreased. The indirect effects were largely a result of increased feed prices for animals, which is mostly corn-based. As corn production was increasingly used for ethanol fuel, the supply of animal feed decreased and caused corn prices to increase. In turn, the feed price increase caused meat prices to rise. In fact, reports have suggested that biofuel production was responsible for a commodity price increase of 10% to 30% and an increase in grain prices of 30% to 40% (Peterson and Wesley, 2009: xix). Therefore, low-income nations and individuals suffered doubly; first with the increase in food prices, and secondly by diverting agricultural land from the production of food to biofuel crops.

The shift in production encouraged by Western governments' subsidy programs for biofuels has also had an impact on land prices. When commodity prices are artificially increased, as has already happened with agricultural subsidy programs, domestic land prices also increase because the returns the subsidies generate are included in assessing the new value of the land (Peterson and Wesley, 2009:146). The simple fact is that subsidies make the rich farmers richer especially when the subsidy is production-based. Wealthy farmers then buy up small farms, increasing farm size and expanding the industrial farm model.

The result is an industrialized agriculture that does not meet the needs of the world's populace in an efficient, safe, sustainable, and equitable way. Industrialized agriculture strips the soil of nutrients and depletes accessible water supplies with its intensive practices. Thus, an increasing amount of hydrocarbon-based inputs are necessary to prevent inherent capacity of farmland from becoming rapidly depleted (Pfeiffer, 2006:2).

The modern, industrial agricultural production performs poorly in comparison with sustainable systems (Pretty, 2005:51) In contrast, sustainable agriculture uses environmental goods and services without harming the ecosystem. Natural processes such as nutrient cycling, soil regeneration, nitrogen fixation, and natural pest control are used, minimizing the use of external inputs that damage the environment and public health. Sustainable agriculture contributes to public goods like clean water, carbon sequestration, flood protection, wildlife, and landscape quality (Pretty, 2005:54). Decentralized sustainable agriculture has been successfully functioning in the United States without any direct and few indirect subsidies.

8.4. THE CASE FOR SUSTAINABLE AGRICULTURE

Agricultural policies should focus on economic development to improve rural incomes and production methods that are both efficient and sustainable (Blank, 2008:440). Sustainable local agricultural operations can be the driver of rural economic development because local farms spend locally. In fact, nonfarm economic activity in rural regions is correlated to the numbers and types of farms (Ikerd, 2008:141) because farmers need to purchase products from local businesses. These purchases create an economic multiplier in the local farm area. As revenues from local farm operations increase, new businesses open, existing businesses strengthen, more local taxes are generated to support local infrastructure, and economic growth follows.

Besides rural economic development occurring, urban regions also benefit because rural people will no longer need to migrate to cities seeking employment opportunities. Thus, cities can also experience less over-crowding and will not have to deal with the pressures associated with urban sprawl. This very likely scenario results from the economic growth that will occur with the expansion of the number of sustainable local farms. This type of economic growth is especially pertinent for developing countries where agriculture is a more vital component of the existing economy and where there is an avalanche of migration from rural to urban areas as people search for employment. Growth in decentralized, local, sustainable agriculture that uses technological advancements and improved farming methods can contribute to macroeconomic growth and stability rather than a country or region being dependent on imported food. However, for this to occur one of the first steps will be the gradual elimination of agricultural and biofuel production subsidies which will improve the competitive position of farms that produce food sustainably.

Skeptics and large agricultural farming operations will scoff at this claim. They will argue that local, sustainable farms will not be able to produce enough food to feed the world. Currently there is enough food produced worldwide, but starvation and malnutrition is a primarily problem of access, due to poverty or geographic location. These access issues result because, like all businesses, agricultural distribution follows the money, whether from government subsidies or wealthy consumers. For example, pineapple and banana plantations run by multinational corporations export their fruit to the developed world because they can command higher prices. Therefore, farmland for producing diverse food products locally is not available because the food is being exported to wealthier regions. Not only are too many people too poor to buy the food locally, but not enough people have the land or financial capabilities to grow their own food (Kimbrell, 2002:33). Moreover, most politicians in poor countries are reluctant to take up the issue of land reform which would redistribute the land to landless peasants for local production rather than for exporting, largely because of corruption.

This book has offered one possible solution to the problem of industrial agriculture; sustainable, community supported agriculture (CSA). The book was written to provide a general overview of agriculture and the general notions of 'sustainability', including the 'CSA model' by using the case study of Roxbury Farm CSA. In Chapter 1, we provided a brief overview of the major issues related to agriculture today and why sustainable agriculture is a necessary solution to these problems. Chapter 2 explored how sustainable agriculture can be used for economic development in both developed and developing countries, as well as to improve food security. Chapter 3 presented the global degradation of soil as one of the most important environmental threats, briefly covered some of the future problems related to agriculture and some different approaches for sustainable agriculture; finally, it introduces some possible solutions. Chapter 4 defined and explored the different sustainable agriculture approaches and methods, highlighting community supported agriculture by providing a theoretical framework for CSA farms. Chapters 5, 6, and 7 developed this theoretical framework, providing both a historical overview and data from the farmers and members of Roxbury Farm in New York. As one of the largest CSA farm operations in the U.S., the case study of this farm illustrates how a farm can be sustainable environmentally, in this case using biodynamic farming methods, and sustainable economically, in the Roxbury Farm case due to large-scale community support.

Together, the history and data of the farm show that the Roxbury Farm CSA is driven by consumer demand, by people seeking relief from the toxic industrial agriculture model and highly-processed food. The history of the farm illustrates the relationship between the farmer and the members. Many of the changes shaping the farm plan that have occurred over the years have been in response to members' comments. The survey data presented in Chapter 6 provide examples of some of the feedback members gave. In turn, members have changed their eating and cooking behavior, as well as their life-styles in response to the seasonal produce they receive from the farm. The members have also responded to help with time, money, and talent when crises have faced the farm. Members of Roxbury Farm CSA are largely concerned about and willing to be engaged in environmental, public health and land use issues.

Not only did people join the CSA because of these concerns, their views are deepened by participating as a member.

These findings are important because they suggest that when consumers learn about the environmental and public health problems associated with industrial agriculture many will demand food produced using sustainable practices. These findings also suggest that consumers who shift their consumption to food produced sustainably understand many of the real negative costs of industrial agriculture. These consumers understand the need for an alternative sustainable agriculture model and will use their income to support their beliefs by, for example, joining a CSA even though there is a high up-front cost in the form of a lump sum membership fee. Members are willing to pay this fee because the money assures farm stability for the growing season and guarantees that the food is grown sustainably. The data also show that consumers of CSA are consciously using their money, time and talents to support local agricultural products and farmers. What is evident from the case study is that both parties, the farmer and the members, benefit from this relationship.

8.5. Thoughts for the Future

This consumer demand and increase in market share for agricultural products produced locally and sustainably has not gone unnoticed by the corporations controlling industrial agriculture. In fact, some organic farming is now becoming industrialized. Industrial agriculture corporations persuaded organic farmers to adopt uniform standards for national organic certification on the premise that uniform standards would provide greater access to new mass markets (Ikerd, 2008:211). However, uniform national and international standards offered the corporations that could meet the minimum standards the opportunity to once again consolidate the market because they could produce and ship large quantities of organic product around the world (Ikerd, 2008:211), thereby reinforcing the current inefficient distribution system. While industrial organic agriculture is a step in the right direction, this approach still depends on growing just a few crops, the application of organic pesticides and fertilizer, and continues to send products very long distances. The result has been a reduction in the number of small, local, diverse organic farms.

We hope this book provides hope that people can move away from the industrial agriculture model that is pervasive in their daily lives. The case study of Roxbury Farm provides an important contribution on how to address the issue of how CSAs can be an alternative to the centralization of the food supply both in the United States and abroad. The status quo of industrial agricultural production is not feasible. What will the change look like when it invariably comes? More local CSA farms would mean that rural economic growth would occur, helping grow healthy local economies; this is especially important in both developing countries where agriculture is often the main economic sector and in the developed world. Industrial agriculture, whether organic or non-organic, is not sustainable environmentally, economically, or culturally. Only a decentralization of food production will create food security, and only agricultural practices that are sustainable environmentally and economically will ensure public and environmental health. Not only does the future of the agricultural sector depend upon the development of local, sustainable agriculture, but quite possibly the future of the humanity and nature as we know it.

References

Armstrong, G.L., Hollingsworth, J., & Morris Jr., J.G.M. (1996). Emerging Foodborne Pathogens: *Escherichia coli* O157:H7 as a Model of Entry of a New Pathogen into the Food Supply of the Developed World. *Epidemiologic Reviews, 18*(1), 29-51.

Blank, S.C. (2008). *The Economics of American Agriculture: Evolution and Global Development.* Armonk, N.Y.: M.E. Sharpe.

Brownell, K.D., Schwartz, M.B., Puhl, R.M., Henderson, K.E., & Harris, J.L. (2009). The Need for Bold Action to Prevent Adolescent Obesity. *Journal of Adolescent Health, 45* Suppl 3, S8-17.

Centers for Disease Control. (2013). Since the 1990s, 45 *Salmonella* Outbreaks Have Been Linked to Live Poultry. CDC CS238230A, http://co.madison.oh.us/health /salmonella-poultry-poster.pdf

Cleeland, B. (2009). The Bovine Spongiform Encephalopathy (BSE) Epidemic in the United Kingdom. International Risk Governance Council, Risk Governance Deficits.

Cummins, J. (2003). Organic Agriculture Helps Fight Cancer. *Institute of Science in Society*. http://www.i-sis.org.uk/oahfc.php

Durden, T. (2014). The Real Inflation Fear – U.S. Food Prices Are Up 19% In 2014. http://www.zerohedge.com/news/2014-03-26/real-inflation-fear-us-food-prices-are-19-2014

Farm Aid. (2013). Corporate Concentration in Agriculture: Limiting Choice and Forcing Family Farmers Out of Business. http://www.farmaid.org/site/c.qlI5IhNVJsE/b.8586841/k.382D/Corporate_Power_in_Agriculture/apps/ka/ct/contactus.asp?c=qlI5IhNVJsE&b=8586841&en=clKNK3NLJbJWJdMOLaKTJaPZImJQKaPWKmKXIdO4LvJdG

Foresight. (2007). *Tackling Obesities: Future Choices*. London: U.K. http://www.foresight.gov.uk/ourwork/activeprojects/obesity /obesity.asp

Hamrick, K.S., Andrews, M., Guthrie, J., Hopkins, D., & McClelland, K. (2011). How Much Time Do Americans Spend on Food? EIB-86, U.S. Department of Agriculture, Economic Research Service, November.

Ikerd, J.E. (2008). *Crisis and Opportunity: Sustainability in American Agriculture*. Lincoln and London: University of Nebraska Press.

Kimbrell, A. (2002). Myth Seven Biotechnology Will Solve the Problems of Industrial Agriculture. In: Kimbrell, A. (Ed.). *The Fatal Harvest Reader: The Tragedy of Industrial Agriculture* (pp. 32-36). Washington: Island Press.

Lappé, A. (2010). *Diet For a Hot Planet: The Climate Crisis at the End of Your Fork and What You Can Do About It*. New York: Bloomsbury.

Marks, R. (2001). Cesspools of Shame: How Factory Farm Lagoons and Sprayfields Threaten Environmental and Public Health. *Natural Resources Defense Council and the Clean Water Network*, July.

McKenney, J. (2002). Artificial Fertility: The Environmental Costs of Industrialized Fertilizers. In: Kimbrell, A. (Ed.). *The Fatal Harvest Reader: The Tragedy of Industrial Agriculture* (pp. 121-129). Washington: Island Press.

Moore, M. (2002). Hidden Dimensions of Damage: Pesticides and Health." In: Kimbrell, A. (Ed.). *The Fatal Harvest Reader: The Tragedy of Industrial Agriculture* (pp. 130-147). Washington: Island Press.

Peterson, E. & Wesley, F. (2009). *A Billion Dollars a Day: The Economics and Politics of Agricultural Subsidies*. West Sussex: John Wiley & Sons Ltd.

Pfeiffer, D.A. (2006). *Eating Fossil Fuels: Oil, Food and the Coming Crisis in Agriculture*. Gabriola Island, B.C.: New Society Publishers.

Pollan, M. (2006). *The Omnivore's Dilemma: A Natural History of Four Meals*. New York: The Penguin Press.

Pretty, J. (2005). Reality Cheques. In Pretty, J. (Ed.). *The Earthscan Reader in Sustainable Agriculture* (2008, pp. 51-63). London: Earthscan.

Wallinga, D. (2010). Agricultural Policy and Childhood Obesity: A Food Systems and Public Health Commentary. *Health Affairs, 29*(3), 405-410.

Wolfe, S. (2014). Why Americans Spend Less of Their Income on Food Than Any Other Country. *GlobalPost*, May 27.

INDEX

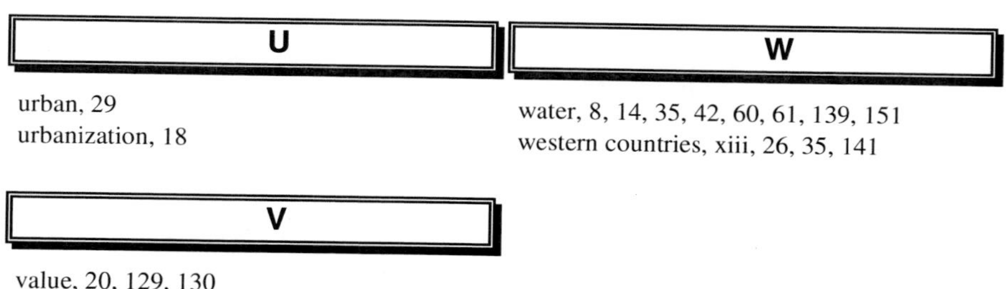
U
W
V